A PHONE OF OUR OWN

A PHONE OF OUR OWN

The Deaf Insurrection Against Ma Bell

Harry G. Lang

Gallaudet University Press
Washington, D.C.

Gallaudet University Press
Washington, DC 20002

All efforts have been made to obtain permission to reprint the photographs
used in this book.

Library of Congress Cataloging-in-Publication Data

Lang, Harry G.
    A phone of our own : the deaf insurrection against Ma Bell / Harry G. Lang.
      p.   cm.
    Includes bibliographical references and index.
    ISBN 1-56368-090-4
    1. American Telephone and Telegraph Company. 2. Telecommunications
devices for the deaf—United States—History. 3. Telephone—United States—
Emergency reporting systems—History. I. Title.

HE8846.A55 L35   2000
362.4′283—dc21   00-028147

Frontispiece: The founding fathers of the acoustic coupler technology that
brought telephone access to deaf people: *(seated)* Robert H. Weitbrecht; *(left)*
Andrew Saks; and *(right)* James C. Marsters. Courtesy of James C. Marsters.

I dedicate this book to all the deaf and hearing people whose countless hours of labor brought us the telephone. To mention you by name would turn the story into a phone book.

CONTENTS

FOREWORD

In the late 1960s, the deaf community made a historical break-through in the world of telecommunications with the introduction of teletypewriters (TTYs) with acoustic couplers. This technology helped us gain access to the regular telephone network some ninety years after Alexander Graham Bell invented the voice telephone. Despite the obstacles that we faced within a monopolistic telecommunications environment, we managed to obtain TTYs from phone companies and volunteer organizations, enabling us to communicate with each other on the telephone infrastructure. The resulting telecommunications movement initially endured a time of skepticism, reluctance, and fear from some members of the deaf community. But that changed as experience over the first few years led us to accept the TTY with enthusiasm, and scores of consumers and professionals began contributing substantially to marketing the technology, developing new products, and improving access. Among those that lent invaluable support were such resource groups as the Telephone Pioneers of America, who played an instrumental role in providing training on the use and maintenance of TTYs.

Little did we realize then how significant this movement would be to the nation's deaf and hard of hearing populations in the areas of empowerment, cohesiveness, and cultural enlightenment. Equally important, the movement has been the foundation for major initiatives in innovation, design, and policy development for today's accessible products and services for the nation's citizens with disabilities, not only in telecommunications, but also in education, employment, recreation, and other life activities.

The TTY movement subsequently led to advanced products and services that now effectively address our informational needs and ensure functional equivalence for us in the mainstream of society. We currently have closed captioning on television and in movie theaters. We have relay services for regular communication with non-TTY users. The 911 centers have TTYs available to respond to emergency calls from deaf people. We work with the Federal Communications Commission (FCC) to ensure that we have full access and opportunity in the telecommunications systems. We are becoming partners with the private sector in the design, development, and implementation processes for new, accessible telecommunications products and services.

A Phone of Our Own describes the events and the individuals that enabled deaf people to enjoy regular communication on the telephone. Robert H. Weitbrecht, James C. Marsters, and Andrew Saks, who together began the first TTY business, Applied Communications Corporation in California, and I. Lee Brody, who established Phone-TTY, Inc., in New Jersey, epitomize the perseverance, resourcefulness, and commitment this development required from deaf people. The book reviews the legislative and regulatory successes from the 1970s to the 1990s that resulted from the tireless advocacy efforts of numerous individuals, such as Karen Peltz Strauss, Alfred Sonnenstrahl, H. Latham Breunig, and Barry Strassler.

The history of the TTY is very much intertwined with the history of another crucial entity—Telecommunications for the Deaf, Inc. (TDI). TDI played a prominent role in the TTY movement, which led to TDI's current stature as a nonprofit, national telecommunications advocacy organization of, by, and for individuals who are deaf or hard of hearing. During TDI's early years, an army of agents was made available across the nation to distribute TTYs and repair them as needed. As the number of TTYs owned by deaf and hard of hearing people grew, the need for a directory of TTY numbers was recognized. The first directory came out in 1968 with 174 listings. Today, the "Blue Book" includes more than 55,000 listings, fax numbers, e-mail addresses, pager numbers, and Web site addresses. TDI advocates to ensure that deaf people have full access and opportunity with mainstream society on the Internet, in the wireless environment, on digital television, and with other technologies that may come our way in the future.

TDI commends Harry Lang for his commitment, vision, and support to produce such an important resource as this book, A *Phone of Our Own*. It is a culmination of thousands of hours of research, solicitation of materials and photos, and face-to-face interviews. Unlike most other book projects where the resources are more readily available, Dr. Lang had to reach out to numerous communities in America and Europe to produce a document that meets the standards of historical research and analysis.

A Phone of Our Own is a celebration of human dignity and personal courage in the life-long struggle for equal access and higher standards of living for the nation's deaf and hard of hearing citizens. May it reinforce or instill in you the responsibility and commitment to build and maintain a brighter telecommunications future for all citizens—regardless of disability—in America and the rest of the world.

Claude L. Stout
Executive Director
Telecommunications for the Deaf, Inc.

ACKNOWLEDGMENTS

I am deeply grateful to Dr. James C. Marsters, one of the three original deaf partners in the Applied Communications Corporation, who provided boxes of historical materials and generous financial support for my research for this book. It was Jim, a true Renaissance man, who had the vision to integrate the different talents of the three original TTY pioneers for the greater good of the deaf community.

Jean Saks, the wife of Andrew Saks, supported my trip to Seattle to study the records of her husband. Her daughter, Andrea Saks, the first hearing pioneer in this history, helped bring the telephone to British deaf people, and she continues to collaborate with others to enhance international telecommunications. In addition to funding my travel to London to examine her papers, Andrea shared many of her experiences as a member of a family that played a pivotal role in this technological breakthrough, as well as discussing her own breakthroughs in Europe.

As Executive Director of Telecommunications for the Deaf, Incorporated (TDI) in 1994, Alfred Sonnenstrahl originally encouraged the organization's Board of Directors to support my research. Both Sonny and his successor, Claude Stout, read the manuscript for accuracy and provided materials and photographs. I am also grateful to the many contributors to TDI's *GA-SK Newsletter*, which was a historical treasure in revealing the significance of this organization to the access movement.

This project was also supported by a grant from the National Science Foundation (NSF) Ethics and Values Studies Program (EVS) under Grant No. SBR-9411871. Rachelle Hollander, the program officer, provided important guidance. Opinions, findings, and conclusions are solely those of the author and do not necessarily reflect the views of any sponsor. The American Telephone and Telegraph Company (AT&T) also provided funds for the project. Sue Decker was helpful with information about AT&T's services for deaf customers, and Michael Catillaz at the National Technical Institute for the Deaf (NTID) was instrumental in obtaining AT&T's support.

I would also like to express my appreciation to John Albertini, Frank Caccamise, George Fellendorf, Gail Kovalik, Marc Marschark, Joseph Slotnick, Kathy Sullivan-Smith, Paul Taylor, Sally Taylor, and Jim Vesper, all of whom provided me with critiques and suggestions. Many other individuals, too numerous to list by name, sent me newspaper clippings, photographs, and anecdotes, or gave me feedback on portions of the manuscript. Special mention should be given, however, to Stephen A. Brenner, Barbara Chertok, Robert Engelke, Jeanne Poremba-Duggan, and Judy Viera for sharing information on the life and work of Robert H. Weitbrecht, the primary developer of the Phonetype acoustic telephone coupler. Jim Haynes and Larry Laitinen, Weitbrecht's longtime ham radio friends, patiently supplied a wealth of information about ham radio, TTY technology, and Weitbrecht's accomplishments. They both provided important support to Weitbrecht during his experiments as well. Anna Terrazzino and Cliff Rowley helped with information about I. Lee Brody. Thanks also to attorneys Karen Peltz Strauss and Sheila Conlon Mentkowski, pioneers in the modern era, who were very helpful in clarifying the various legislative actions in the 1980s and 1990s that led to the Americans with Disabilities Act of 1990, especially Title IV and the nationwide telephone relay system. Senator Tom Harkin also shared his experiences with me.

Gail Kovalik, Judy DeWitt, and Linda Coppola at the Rochester Institute of Technology, Michael Olsen and Ulf Hedberg at the Gallaudet University Merrill Learning Center, and Judith Anderson at the Volta Bureau provided ongoing assistance with my searches for library information and materials. Mary Ann Erickson and Gayle Meegan assisted with correspondence and various clerical assignments. Mark Benjamin, Bruce Letzelter, Darundee Sa-areddee, Kitae Kim, and Hon Chung Siu spent many hours scanning the photographs that help tell this story. Their magic touch on the computer gave new life to many of the faded pictures. I am delighted with the editorial commitment and care of John Van Cleve and Ivey Pittle Wallace in helping me tell a story of such importance to the life of the deaf community.

Finally, I am deeply indebted to my wife and colleague, Bonnie Meath-Lang, who shares with me the joys of communicating in many ways. With love and gratitude, I thank her for her support and patience during the five years of research and writing of this book.

To tease out the historical interaction in telephony between what was technically possible, economically profitable, and socially desirable is a puzzle worthy of scholarship.

—Ithiel de Sola Pool, *The Social Impact of the Telephone*

INTRODUCTION

For nearly a century after the advent of the voice telephone, we deaf people were without a phone of our own. We had to carefully plan visits, vacations, and business transactions. Conducting even basic daily exchanges was a difficult chore. A last-minute change in plans for a meeting, for example, presented special problems. Hearing persons could not always be found to help make telephone calls, and when deaf people missed engagements, there was concern among those who were expecting them. It was very common for deaf people to use letters and postcards or to drive long distances for face-to-face communication, often finding no one at home.

Today there are hundreds of thousands of telephone devices used by deaf people, and a variety of services available to them for personal, social, and professional communication around the world. Telephone access is now available, for example, not only through the TTY but also through computers with TTY-compatible software, which allows deaf people to make telephone calls directly on their computer. The TTY has opened opportunities for us to pursue careers previously not available. Through telephone relay services, we can call anyone by having our typed messages relayed in voice and by having voiced messages typed to us. For all its simplicity, the TTY has led to extraordinary changes in deaf people's lives. It has created links in the community, the indispensable communication needed to sustain relationships and build new ones. It has given us increased flexibility by making it easier to purchase goods, book reservations, schedule appointments, seek medical advice, and solve crises. The TTY has brought new educational opportunities, along with opportunities to internalize the values, norms, rules, and symbols of the expanded community through long-distance communication. Most importantly, it also has provided comfort and safety.

Long-distance communication is essential for participating in society, whether for mundane chores such as calling an automobile repair shop or for more important interactions such as telephoning a teacher about the progress of a child in school. Such communication plays an important role in determining the extent to which deaf people become integrated into daily community activities. Prior to the development of TTY technology, the daily dependence on neighbors, friends, and family members for assistance in making calls frustrated many deaf people. As one deaf person wrote in February 1935, "when someone with normal hearing is not at home, no one is as far as telephone calls are concerned. This is just one more case where we must depend upon our long-suffering families for assistance."[1]

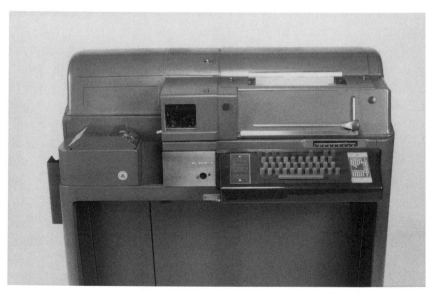

The Teletype Corporation Model 28 ASR teletypewriter was a first generation TTY. It weighed 260 pounds and measured 39" × 36" × 18". From the collection of I. Lee Brody, courtesy of NY-NJ Phone-TTY, Inc.; photograph by George Potanovic, Jr./Sun Studios.

When a Phonetype acoustic coupler was connected to the TTY, deaf people could place and receive local and long-distance calls over regular telephone lines. Courtesy of James C. Marsters.

There is no better term than "behemoths" to describe the first tele-typewriters (TTYs) deaf people used to make phone calls. The enormous, old, and heavy machines were the size of big drop mailboxes. They weighed several hundred pounds and stood more than four feet high.[2] Appropriately, many were painted battleship gray. Vibrations from the TTY's gears and motors shook the floor and penetrated walls. The rumbling could be heard by neighbors living in adjacent apartments. Inside the huge machines were vibrating levers, rotating parts, shafts, clutches, pawls, plates, springs, and screws. Electric wires connected magnets, transformers, and capacitors. With meshing gears and slipping clutches, the behemoths spewed forth heat and sometimes sparks.

The teletypewriter (or teleprinter) alone, however, was not enough to make our telephone. We also needed to transmit and receive electric signals over the telephone line. These signals needed a modem (also called a "coupler") to convert them to pulses that could activate the teleprinter keyboard. In 1964, the deaf physicist Robert H. Weitbrecht developed such a coupler. Along with James C. Marsters and Andrew Saks, Weitbrecht marketed the coupler under the name "Phonetype." My own first telephone was a combination of the new technology (Weitbrecht's Phonetype modem) and recycled older technology, the teletypewriter.

Primitive as they were, our first TTYs filled a void of loneliness and solitude. We found the old machines in surplus stores, scrap heaps, and corporate warehouses, and they became cherished treasures that we cleaned and displayed prominently in our homes. Often, they sat in kitchens where lighting was best. With the new coupler and a teletypewriter, we broke down the great communication barrier that had isolated us from each other as well as from our hearing friends.

The reminiscences of Sylvia Schechter from Brooklyn, New York, typify our struggles before the TTY, Weibrecht's coupler, and the relay service changed things. "I can't begin to tell you how the TTY has improved my life," she wrote to me. Her daughter had lived upstairs in a two-family house and received and made all her calls. But when her daughter moved to Nashville, Tennessee, there was no one to make the calls for her. "To make a phone call, I would have to go across the street to ask a neighbor to please place a call. You can't imagine how humiliating it was and how depressing." When Mrs. Schechter learned about the TTY, her daughter and sister both bought one too, but they were the only people she could call. "All my hearing friends, my doctor, my dentist, I couldn't call. I lived in Brooklyn and my sister lived in Wardmere but I'd call her to ask her to make a call for me. It wasn't easy—life was very hard."[3]

There were elements of both comedy and tragedy in the ninety years of telephonic silence. In 1941, as the United States was preparing for possible war with Germany, Shirley Fischer (Panara) attempted to communicate with a deaf friend by tapping a pencil on the mouthpiece of the telephone handset. She was able to identify the pre-established code with the help of a hearing aid. One night in August, two well-dressed gentlemen visited the family residence to inquire about the mysterious phone conversations. The agents asked the young deaf woman to verify that she was not involved in espionage.

Several decades later, the telephone remained a useless piece of plastic and metal for most deaf people, a situation that was especially noticeable during an emergency. When a deaf man in New Jersey had a heart attack early one morning in the late 1960s, his wife frantically sought help. She rushed out of their home screaming, but her frightened neighbors kept their doors shut. By the time she spotted a squad car on a patrol and asked the officer to make the emergency telephone call, it was too late — her husband had died. When another deaf man in Eau Claire, Wisconsin, suffered a heart attack and had no way to alert anyone, he drove himself to the hospital, where he died within a few hours. These and many similar experiences showed a critical need for assistance from telephone companies.

Such help was not forthcoming for a long time, however. In the 1960s, one telephone company official actually recommended that deaf people shout repeatedly into the telephone or leave the phone off the hook during an emergency in the hope that the operator would notice and dispatch the police.[4]

The value of the telephone in an emergency was brought home to me in a powerful personal experience. One evening while I was working on this book, my wife Bonnie ran into my study to tell me that something had happened to Michael Thomas, a deaf friend, while she was talking with him on the telephone. We rushed to Michael's home and found him bent over in excruciating pain. Immediately, we took him to a hospital, where we stayed with him through emergency surgery. In the waiting room, I thought of the many reports I had read about how TTY and voice telephone calls had saved lives. But until that evening, I had never had such an experience.

A *Phone of Our Own* begins with the original conception of the "telephone for the deaf" in 1964 and ends with the signing of the Americans with Disabilities Act in 1990. The narrative is based largely on original correspondence and documents loaned to me by families and friends of the primary actors in this history. The documentation includes brittle, faded thirty-year-old printouts deaf people had saved from precious conversations that took place long ago. These records, in particular, led me to a fuller understanding of the dedicated efforts of Weitbrecht, Marsters, and Saks and the community that followed in their footsteps in the years after the TTY modem's development.

Weitbrecht, Marsters, and Saks were remarkable pioneers. From the outset, they established four goals for telecommunications access for deaf people: availability, affordability, portability, and accessibility. The first three goals were pursued almost quixotically at a time when mainstream society and corporate culture, especially, stood immobile, like giant windmills refusing to be moved. By 1966, the deaf partners had developed the first TTY modem for deaf people, thus creating an available and (with mass production) affordable telecommunications device and realizing the first two goals. They also began experimenting with making such a device portable.

The fourth goal, accessibility through a national telephone relay system, would require a change in the political culture that Weitbrecht, Marsters, and Saks could not affect by themselves. Even in this matter, though, these deaf pioneers were important, because they were the first activists to gain government recognition of our particular telecommunications needs. They also identified the necessity for long-distance rate reduction for deaf people because of the time required to type phone messages, and they challenged the Internal Revenue Service to obtain a ruling that allowed the cost of the telephone coupler to be deducted as a medical expense.

Their legacy of self-advocacy, along with the hard work of many other deaf and hearing people, evolved into a social and technological revolution in the 1970s. As former National Association of the Deaf President Robert G. Sanderson said in a letter to me, "Those guys will go down in history as making perhaps the most significant advancement of deaf people's welfare since the first school for the deaf was established. We all—every deaf and hard of hearing person who ever touched a TTY—owe them a debt of gratitude that is beyond payment."[5]

Within the deaf community, Weitbrecht has been honored for his development of the telephone coupler, often referred to as the

"Weitbrecht modem," but the role of Marsters and Saks is less well-known. They provided Weitbrecht with guidance, significant financial and personal support, and vision. The volumes of letters and memoranda dealing with the implementation of TTY technology amassed by Weitbrecht include hundreds of memos from his two partners that focus on coupler design and other issues. The obsession of these three men over breaking down a major communication barrier consumed their lives. Each man brought unique skills to their partnership, which ultimately was less of a business enterprise than a philanthropic adventure.

The involvement of other members of America's widespread deaf community is also absent from most accounts of the struggle to obtain telecommunications access. During the early years of TTY development, small cadres of deaf volunteers around the country reconditioned discarded teletypewriters and established local TTY networks. Many were blue-collar workers unknown outside their immediate circle of deaf friends. This book sets out to make these courageous people visible and to demonstrate the deaf self-advocacy that led to the TTY's eventual success.

In telling the story of these pioneers, I could have used many possible approaches. For example, I could have used a predominantly technical approach, emphasizing the engineering challenges involved in the invention of the TTY, or I could have focused on the politics of deafness and its relationship to technology and society. I have chosen instead to blend technical and sociopolitical analyses into a description of the work of Weitbrecht, Marsters, and Saks over several decades while they witnessed and participated in the astounding impact of the Phonetype modem. The book summarizes a long and difficult struggle for fair and impartial telephone services and products — in reality, it was an insurrection by the deaf community against the telephone industry that was brought on by personal, corporate, societal, and other value systems in conflict with one another. Furthermore, it is critical that future generations of minority groups and people with other disabilities know the story of the TTY modem. Its development marked the beginning of a new era of self-advocacy. And as technology plays an ever-increasing role in the lives of all people, it is even more essential that all of us participate in the decisions that may influence the way we live and work.

The great eighteenth-century philosopher Jean-Jacques Rousseau, who experienced deafness himself, once came upon an immense Roman aqueduct and remarked that the engineering feats not only reminded him of his own smallness, but elevated his soul.[6] Rousseau, a powerful proponent of human rights, would never have imagined the feat of engineering and social rights deaf people enjoy today in the form of accessible telecommunications. But, as A *Phone of Our Own* reveals, these rights were not easy in coming.

1

A CHANCE ENCOUNTER

The telephone companies have not offered anything at realistic rates to deaf people, so some of us had to "go at it" on our own to develop a suitable communication device using . . . cast-off teleprinters.

—Robert H. Weitbrecht, letter to Freeman Lang, September 15, 1966

On April 11, 1963, deaf physicist Robert Haig Weitbrecht turned forty-three years old. He was living in the hills west of Redwood City, California, in a new two-bedroom duplex on Woodside Road. Weitbrecht had converted a bedroom into a radio "ham shack," and his living room was strewn with radio equipment, electrical meters, boxes of electronic parts, and books. Scattered around his bedroom were issues of *RTTY Journal*, a periodical for radioteletype users. A few months after his birthday, Weitbrecht's chance encounter with the father of a deaf child would change deaf people's lives forever.

Weitbrecht was an unlikely person to become a hero for the American deaf community. For much of his life, he had stayed apart from deaf people, socializing with them infrequently, perhaps because of an overprotective mother and memories of childhood teasing about his deafness. But his parents had nurtured in him a love of science, and this fascination was compelling throughout his life. It was also essential to his success in developing the telephone acoustic coupler.

As a child, Weitbrecht developed a special interest in astronomy and receiving Morse code signals by feeling vibrations. At fifteen, he was allowed to connect his own practice oscillator—complete with batteries and a headphone—to the Federal Communications Commission (FCC) code-sending machine to demonstrate that he could receive Morse code at thirteen words per minute, the time necessary to be eligible for an amateur radio license. A month later, his mother interrupted his class and hand-delivered the license as his classmates looked on. He was officially a "ham."

Weitbrecht's interest in science guided both his choice of professions and his hobbies. He began his college career at Santa Ana Junior College in 1938 and then moved on to the University of California at Berkeley, where he received his bachelor's degree in astronomy with honors in 1942. He worked as a physicist at the Radiation Laboratory at the University of California and as an electronics scientist with the Manhattan Project during the war. In 1949, he was honored with the Superior Accomplishment Award by the U.S. Naval Air Missile Test Center in Point Mugu, California.

Morse code transmitted by radio waves became Weitbrecht's particular obsession because it permitted him to communicate with other radio hams despite his deafness. In 1950, he sought to expand this long-distance contact by acquiring a used Model 12 "receive only" teletypewriter, usually called a TTY (and sometimes called a teleprinter),[1] from a Los Angeles newspaper plant. With the new machine, he could receive radioteletype communications from Japan, the Philippine Islands, Australia, South America, and many places in the United States.

Before long, though, he realized that receiving radioteletype messages was not enough to satisfy him. He also wanted to send his own. He searched for six months and finally procured a keyboard from an East Coast ham. Using a string around the gear and the shaft of an old washing machine motor, he managed to adjust the speed until the mechanical keyboard worked. He had his first "send and receive" teleprinter outfit. This was the first time Weitbrecht had full visual access to long-distance radio communications. Prophetically, he wrote "RTTY . . . is now an important and growing facet of Amateur Radio, with untold possibilities for communications purposes."[2]

Weitbrecht's experience with his radioteletype station, one of the first on the air since amateur radio began, taught him the value of challenging government communications regulations. In January of 1951, he petitioned the FCC to permit radioteletype operation on a broader range of frequencies. After serious consideration, the FCC granted his request. The successful challenge opened more opportunities for RTTY communication among his amateur radio friends.

Weitbrecht left California in 1951 and moved to Yerkes Observatory at Williams Bay, Wisconsin, where he became known to other hams as "The Wisconsin Wizard." At Yerkes he designed electronic instrumentation for use in astronomical research, earned a master's degree in astronomy from the University of Chicago, and aided in the development of the worldwide WWV-WWVH Radio Time Signal adopted by the National Bureau of Standards. He took many trips with radio users, lugging around huge pieces of equipment. During one trip in 1957 with his friend, Bruce Rowland, Weitbrecht drove 600 miles with a Model 15 TTY in his station wagon. Early one morning, he woke Bruce up to catch "Sputnik" speeding across the sky.

Robert H. Weitbrecht in his ham radio shack in his home in Redwood City, California. His FCC call sign was W6NRM. Courtesy of James C. Marsters.

Robert H. Weitbrecht on Lassen Peak in 1963. With him is his faithful companion Blackie. Courtesy of Laurence H. Laitinen, W7JYJ.

Weitbrecht's travel with bulky radio equipment and heavy teletype-writers in the 1950s probably shaped his later belief that equipment portability was essential for deaf access to telephony. Since he could not hear on the telephone, he took along his radioteletype gear to communicate with his friends. The teletypewriter, with messages transmitted via radio waves, replaced the voice telephone for his long-distance contacts. Articles in *RTTY* from this period include reminiscences of a 4,500-mile trip he took through a dozen states to visit fellow hams, as well as technical reports on his experiments with electronic transmitting distributors, bandpass filters, and transistor designs. Weitbrecht became such an ardent experimenter with various teletypewriters that he soon earned a new nickname from his friends, "Mr. Terminal Unit." The 1954 issue of *RTTY* included a photograph of Weitbrecht with a Model 26 TTY and his dog Blackie.

The radio equipment Weitbrecht carried around the country in his Chevrolet station wagon was so heavy that he began to search for ways to make things more compact. In 1958, Weitbrecht left Yerkes and returned to California, this time to work at the Stanford Research Institute. In fact, he reported in *RTTY* about a trip to the San Mateo, California, Hamfest in 1959. "It is always interesting to see 'how portable' one can make his station, no matter whether high powered or low powered."[3]

He also sought ways to win others over to radioteletype technology. "Sure a lot of effort and work to move and set up gear, but it's all in fun and we wanted to demonstrate RTTY to the 'unwashed multitude'. . . . Maybe we'll have a few more converts, yet!"[4] In another article, he sounded off his anger over the smashing of used TTYs by telephone companies (to assure that the machines would never be used in commercial competition). Pacific Telephone Company had recently released some machines to radio amateurs, but he noted with regret that amateur use of teleprinters would never grow with the destruction of TTYs by many other telephone companies. "We are sure AT&T Co. and its subsidiaries want us to win the next war, but a little more cooperation in making available the cast off machines would be a great help in the amateur radio self-training field."[5] These early struggles with portability and availability of TTY equipment prepared him well for his later battles in support of the deaf community's quest for telephone access.

The story of the "telephone for the deaf" began on a summer day in 1963 when Weitbrecht was hiking on Lassen Peak with his young hearing friend Larry Laitinen. He and Laitinen shared a common interest in photography.[6] Since early 1960, they had enjoyed excursions with their cameras at Lassen Peak, their favorite stomping grounds. Lassen Peak is one of the few active volcanoes in the United States outside of Alaska and Hawaii. Weitbrecht loved the mountains. Climbing Lassen Peak brought him closer to the stars. During his treks he also dreamed of one day flying above the clouds in his own plane.

While pausing near the top to take in the view, Weitbrecht and Laitinen met Edwin McKeown, a lawyer, and his eleven-year-old deaf son, Laddie. McKeown had heard Weitbrecht speaking to Laitinen and recognized the distinct voice quality of a person born deaf. He introduced himself and Laddie to the physicist, explaining that his wife, Patsy, was also deaf. McKeown told Weitbrecht that Laddie was in a public school. The boy was doing all right, but he was challenged by his deafness.

This reminded Weitbrecht of his own childhood. Born deaf in 1920 in Orange, California, Weitbrecht was brought up with his younger brother George by their mother. His father, a farmer, lost the family farm during the Depression and died when Weitbrecht was young. After being tutored at home for a while, Bob entered a public school. He performed at the same level as his hearing classmates but was often teased and treated cruelly because of his hearing loss. The experience left him somewhat bitter and to his last day, he struggled with self-confidence. He hoped Laddie would not experience similar treatment. After talking a little longer with the McKeowns, Weitbrecht and Laitinen went on their way.

In February of 1964, Weitbrecht heard from McKeown again, this time in Piedmont, California. McKeown invited him to a dinner party with some deaf friends. Weitbrecht had not associated much with other deaf people. When he was four years old, a retired teacher had tutored him and another profoundly deaf child. Later, while studying astronomy in college, he had befriended a deaf student, who taught him sign language. But he never had much opportunity to use it because he preferred to read lips. This had made him rather uncomfortable about attending the McKeown dinner. At the party, however, Weitbrecht learned that the McKeowns' deaf friends were all established professionals who, like himself, could read lips well. They enjoyed his fascinating accounts of working with electronics instrumentation design. There were few deaf scientists in the workforce at this time and that made his accomplishments all the more remarkable to them.

One of the professionals at the dinner party was Arthur Simon, a deaf book editor. Within a short time, Simon contacted Dr. James C. Marsters in Pasadena to tell him about Weitbrecht. Marsters at the time was an orthodontist with his own office and a full clientele. He communicated with his hearing patients by reading their lips. When this was not possible, his dental assistant repeated their words. Like other deaf people, Marsters found ways around most communication barriers, but he had never found an adequate solution to the problem of telephone access, despite more than two decades of searching for a way to use the common household telephone. When he learned of Weitbrecht's electronics background, he felt that destiny had brought them together.

Born in 1924 in Norwich, New York, Marsters became deaf as a result of maternal rubella. He attended elementary school in Norwich and worked with a tutor after school. He graduated from the Wright Oral School for the Deaf in New York City in 1943. During World War II, he entered an accelerated wartime educational program at Union College in Schenectady, New York. After graduating from Union College in 1947, he moved to New York City. A year later, he began applying to dental schools.

Because of his deafness, colleges of dentistry repeatedly turned him down until, after two years of persistence, New York University finally admitted him. He successfully completed the College of Dentistry program and passed the dental licensing examinations for the states of New York and California. In a sense, Marsters' struggle for admission to dental school prepared him for the long battle ahead to gain access to the telephone.

By the time Marsters heard of Weitbrecht, he had already fought and won a skirmish with the telephone company. He had sought permission to use a Listening Head, a device that allowed him to speak into the telephone as he read the lips of a hearing person who was repeating the voiced message to him. When he first connected this device directly to the telephone's main base, he was warned by the telephone company that the connection was illegal. He had to struggle for years before gaining permission to use the equipment.

Marsters had also experimented with other ways to try to gain access to the telephone. He tried a portable battery-operated amplifier, a loudspeaker, and a magnetic pick-up coil, which, when held near the earpiece of a telephone receiver, worked well enough to at least let him know that someone was saying "no," "yes-yes," or "please repeat" as he talked on the telephone. Marsters wondered if Weitbrecht

Dr. James C. Marsters with a leased Beachcraft Bonanza. Marsters and Weitbrecht shared a love of flying. Courtesy of James C. Marsters.

could help by developing a better telephone device. The fact that Weitbrecht was deaf motivated Marsters to contact him even more. At the time, Marsters and Andrew Saks, a deaf engineer, were searching for ways to inspire deaf youth to consider scientific careers. In April 1964, the deaf orthodontist wrote to Weitbrecht, hoping to recruit him for this endeavor.

Weitbrecht's ham radio hobby made Marsters curious, too. "I am much interested in getting such a system, if practical, set-up in my own home," he wrote. "I know it involves getting a radio ham license, etc., but it doesn't faze me."[7] Marsters proposed a first meeting on Weitbrecht's birthday, April 11. He also invited Weitbrecht to go with him to Voice Systems in Campbell, California, to see a "telephone gadget" the company had developed. Marsters had made many of his own doorbell and alarm clock signalers and was curious about what the company had to offer.

Marsters piloted an airplane from Pasadena to San Francisco to visit Weitbrecht, who welcomed him into his Redwood City home. Within a short time, Weitbrecht showed Marsters his RTTY system. Marsters was intrigued by the amateur radio set-up in Weitbrecht's cluttered two-bedroom duplex. At the same time, Weitbrecht was fascinated that Marsters had piloted a plane to visit him. The conversation jumped back and forth between radio and flying as their friendship took root.

Something else in the room caught Marsters' attention. It was a TTY connected to a private telephone line. In the early days of amateur RTTY, "on-the-air" roundtable discussions were popular. Groups of friends would gather almost nightly on a particular frequency. After making some remarks, each operator would indicate which station was going to transmit next and then turn the send-receive switch to receive. It was also customary to allow a slight pause before transmitting in case some newcomer wanted to join the discussion. Weitbrecht explained to Marsters that he was experimenting with attaching a switch to the teleprinter keyboard that would turn the transmitter on when any key was struck. This would make such conversation easier by eliminating the separate send and receive steps. The sending operator could type as much as he wished, then simply pause to allow the transmitter to automatically turn off so that someone else could send. Weitbrecht and his hearing friends within the San Francisco Bay area also conversed over TTYs connected directly to the private telephone line. He had been using both ham radio and the telephone for some time, and he showed Marsters some of the printed phone conversations.

Marsters saw in the private telephone line setup the germ of an invention that might fulfill his dream of more than twenty years. After returning home to Pasadena, he thought a great deal about it. If two people could communicate with TTYs through radio transmission or over the telephone lines in a local circuit, then why could they not do so over *any* telephone line? Regular radioteletype might be successful for some, but there were too many deaf people who did not have the time, money, or interest to acquire a radio license. He wrote to Weitbrecht on April 26, 1964, and planted the idea of a national network of TTYs for deaf people — *one that would also be designed by deaf people:* "What I have in mind, Bob, is the possibility of a network of regular telephone line RTTY for deaf people who can afford one. . . . but not to lease one via the telephone company nor a special telephone line. . . . why won't it be possible to translate over the regular line by proper modification of equipment. . . . granted that eventually there will be enough units? . . . What do you think?"[8]

Weitbrecht agreed with Marsters about the potential of extending TTY communication so that other deaf people might use the telephone on a regular basis. He responded to his new friend immediately. "About telephone setups for the deaf using teleprinters. I see no reason why not. I have such a deal setup now operating through my private line telephone. Nothing special about the telephone itself; I merely built an adapter unit that I hook up to the line; it converts teletype signals into tones that go readily through the landline; likewise incoming tones are converted into current pulses to run the teletype machine. . . . In fact I am now working on designs for such 'phone patch teletype' towards a simpler and more effective teleprinter system for working over regular telephones — any telephone; not just a special teleprinter line."[9] Weitbrecht knew that the patch units, directly connected to the telephone lines, were frowned upon by telephone companies because of their desire to restrict connections of private equipment to the phone lines. For this reason, he had used patch units cautiously, hoping that some day the companies would change their policies.

Marsters was convinced that an exciting venture was ahead. Many deaf people would view the TTY as a major breakthrough in their lives. He wrote to Weitbrecht in Redwood City, "Will be glad to be your first customer for a phone-patch teletype."[10]

Marsters also introduced Andrew Saks to Weitbrecht through his first letter in April: "Andy might be much interested in your radio-ham gear and other gadgets, too, as well as you as a person."[11] In May of 1964, Weitbrecht met Saks at another dinner party. Saks and Marsters

Andrew Saks in his office in Belmont, California. Courtesy of Jean M. Saks.

had been friends since the early 1950s and had shared a common interest in business investments over the years. Although he was indeed interested in the telephone access issue, Saks was very cautious about committing to making an investment. He had recently sued a stockbroker and lost the trial. Marsters had to convince Saks that the idea of supporting Weitbrecht's research and development efforts was worth further consideration.

Born in 1917, Saks was deafened by a mastoid infection when he was eight months old. The grandson of Andrew Saks, founder of the New York City Saks Fifth Avenue department store, he attended Lafayette College in Easton, Pennsylvania, and then studied electrical engineering at the University of California at Los Angeles. In 1941, while a student, Saks authored an essay titled "Deaf Difficulties," in which he wrote about the problem of not being able to use the radio or the telephone. During World War II, he held various positions for North American Aviation, the Electrical Products Division of Western Electric Company, and the Douglas Aircraft Company. After the war, he conducted research in blood physiology for eight years at Stanford University as part of the nation's space program. Later, he managed his own investment business.

Like Marsters, Saks had tinkered with visual communication devices that would assist deaf people. He had worked on relay coils and flashing light signalers to let him and his friends know that the telephone was ringing or that someone was at the door. He also had worked on an early version of a signaler that would allow deaf parents to know a baby was crying. His drawers and closets were filled with various gadgets, and his wife, Jean, was never sure what he would bring home next. He was also experienced in the business world, which would be helpful to his new partners.

Saks, Weitbrecht, and Marsters shared an interest in mechanical devices and much more. All three were deaf professionals at a time when such accomplishment was comparatively rare, in part because of the limitations imposed by telephone inaccessibility. They also were independent and believed that deaf people could and should help themselves instead of relying on hearing people. Each was, to some extent, frustrated by the lack of easy phone access. Each man would play an important role in ending this frustration for hundreds of thousands of other deaf people.

2

UP THE MOUNTAINSIDE

After numerous conversations with Marsters and Saks, Weitbrecht evaluated the general obstacles he faced in developing a telephone device that deaf people could use—one based on visual communication. He believed that an acoustic coupler used with the TTY was probably the best idea to pursue, but first he had to review other available equipment. Second, he would have to work with Marsters and Saks to develop strategies to connect deaf people to some sort of network once the equipment became available. This would not be an easy task. For one thing, voice telephone technology was improving rapidly. Therefore, a constant effort would be needed to keep abreast of new advances so that the equipment for deaf people would not become obsolete too quickly. Third, Weitbrecht needed to find a solution to the problem of calling hearing people who did not have a visual telephone device. Some kind of service would be necessary to link hearing and deaf telephone callers who used different equipment. Without an intermediary to "relay" voiced and teletyped messages, the network of deaf people would be rather limited.

Unfortunately, the ninety-year wait for a visual form of the telephone for deaf people had not been necessary. Accessible telephone technology for deaf people was available earlier. On February 14, 1876, both Alexander Graham Bell and Elisha Gray filed for a patent on the voice telephone. Bell filed his claim two hours earlier than did Gray, and he successfully defended his patent against lawsuits that followed. There is a double irony to this story. The first is that Bell's primary interest was teaching deaf children. Supported by Thomas Sanders and Gardiner Greene Hubbard, whose deaf children he had been instructing, Bell labored long hours with manometric gas flames and

21

other instrumentation to "render visible to the eye of the deaf the vibrations of the air that affect our ears as sounds."[1] At the same time, he was experimenting with the harmonic telegraph, which distinguished musical notes and made it possible to transmit several messages over one wire simultaneously. His interest in acoustics and telegraphy eventually led him to the idea of the telephone, an invention that was useless to deaf people.

The second irony was that Elisha Gray, the man who lost the patent to Bell, was the first person to develop a telephone device that deaf people could use. Although not intended specifically for deaf people, Gray's "Telautograph," introduced at the 1893 World's Fair in Chicago, met the needs of deaf people. Messages handwritten at one end of a wire with a pen-lifting mechanism were reproduced automatically on the other end through the use of a stylus and a wide sheet of paper. The Telautograph created a sensation in deaf communities in several countries. "It seems to us," wrote the editor of a magazine for deaf people in Derby, England, "that not only will this new invention be a success, but it will be an especial boon to the deaf. . . . a *deaf* person can use the instrument with the same advantage as a hearing person."[2]

But hearing people controlled the telephone industry, and they had grown accustomed to the voice telephone. As the American deaf community's periodical, *The Silent Worker*, commented in April 1900: "Gray's telautograph would be the very thing for the deaf if used in connection with the telephone. . . . When first brought out a test of the system was made . . . but while it worked all right it was found that business men disliked the bother of writing. . . . Its failure to 'take' with the public is cause for regret to the deaf to whose needs it is so well adapted."[3]

Long before Weitbrecht, several other deaf inventors tried to develop other visual telephone devices. William E. Shaw of Lynn, Massachusetts, was one of the most prolific deaf inventors, credited with at least fifty inventions specifically for deaf people. His "Talkless Telephone," also called the "Deaf-Mutes' Telephone," was a light board with thirty-six incandescent lamps that had letters and figures painted on the head of the bulbs. Each deaf person on the telephone had a

The Telautograph, a telephone device created by Elisha Gray, was first demonstrated at the World's Fair in Chicago, 1893. Courtesy of Danka/Omnifax.

keyboard connected to these rows of electric lights that transmitted and received messages, letter by letter. In 1912, the *Technical World Magazine* reported that

> the new phone represents the work of a man and his wife . . . assisted by their young son, who is almost blind. . . . The deaf [person] who wishes to talk with another person presses the keys in order, spelling out the words as on a typewriter, his vis-a-vis reading off the letters as they flash up on the lamps. The keys come down on points of contact in the same manner as do the printing typewriter telegraph machines, instead of striking by means of a lever bar as does a typewriter key. This does away with any false or lost motion and ensures perfect contacts. . . . the letters can be read as quickly as they can be flashed up. Thus persons familiar with an ordinary universal typewriter keyboard could readily operate the machine, and with a little practice become expert at it.[4]

In 1964, Weitbrecht did not know about these very early failed telephone development efforts. He did know, however, about the Electrowriter, which operated much like Gray's Telautograph. The San Fernando Valley State College Leadership Training Program in Northridge, California, was conducting telephone communication ex-

Deaf inventor William E. Shaw and his wife standing before his "Talkless Telephone," which he developed in 1912. *Boston Daily Evening Item*, March 12, 1912.

periments to find ways deaf leaders could use the telephone. Begun in 1961 and funded by the Rehabilitation Services Administration, the Leadership Training Program provided graduate training to established deaf professionals working in the deaf community who wished to become administrators. Like other participants in the program, Henning Irgens, a deaf consultant from the Michigan Rehabilitation Institute, desperately needed telephone access. Irgens compared the plight of deaf people with the legendary Sisyphus. For saving people from the brink of death, Sisyphus had earned the wrath of Hades. The god Zeus was also angry and punished him by denying him freedom. Sisyphus was forced to start his labor over again each time he appeared near success. Irgens used the legend to describe the social and professional barriers the telephone had presented deaf persons. "Nearly all deaf people still have to go to great lengths to spend money for travel and to indulge in exhaustive correspondence in order to accomplish designed tasks which otherwise might easily have been solved over the telephone. . . . [The telephone] has subjected many deaf persons to certain indignities in relying upon hearing persons for calls."[5]

Writing machines such as the Electrowriter, also known as "tele-pens" or "remote scratch-boards," allowed callers to communicate through handwritten messages without the use of speech or hearing. While a January 1964 demonstration of the Electrowriter in North-ridge had generated enthusiasm, some reports exaggerated the electric writing machine's potential. The President's Committee on Employ-ment of the Handicapped extolled the device: "Its merits for ending the isolation of the deaf are so apparent, a general acceptance should await its arrival."[6] That was far from the case. An Electrowriter cost more than $1,200, and the legibility of the messages depended on the clarity of the individual's handwriting. Use of this type of telephone device took a much longer time, increasing costs for long-distance communication. In addition, it was necessary to have an identical brand on both ends of the telephone line, introducing a problem for marketing the technology.

Irgens was unaware that Weitbrecht was undertaking the Sisyphean task of developing a working telephone modem. Furthermore, Weit-brecht was working not far from Northridge, where the Leadership Training Program was located. He kept a watchful eye on *The Deaf American* magazine for summaries of the telephone gadgets being used by the Leadership Training Program participants. One was the "Speech Indicator," developed by an electronics education specialist for the Los Angeles City Schools.[7] It included an indicator needle that was deflected by sound. A short burst of sound such as the word *no* spoken on the phone would move the needle once. Two sounds such as *yes-yes* moved the needle twice. A third code, *please repeat*, moved the needle three times. At the Leadership Training Program, deaf people with intelligible speech were trained to do all of the talking on the phone while watching the Speech Indicator for *no, yes-yes,* and *please repeat* responses.

But Weitbrecht himself could not use the Speech Indicator com-fortably. His speech was not clear enough to be understood over the

The Speech Indicator was developed by Hugh Moore for use by students in the Leadership Training Program at San Fernando Valley State College. The needle moved when a person on the other end of the line spoke into the telephone. Courtesy of Alfred Sonnenstrahl. Photograph by Mark Benjamin/National Technical Institute for the Deaf.

telephone. Marsters' speech was easier to understand. He was taking private speech lessons with the Countess Elektra Rozanska, who was a private tutor to Hollywood actors as well. Weitbrecht's friend, Patsy McKeown, also used the Speech Indicator, but with a different code. As she spoke into the telephone, she required the person on the other end of the line to dial a four, then a one, to produce a "dah-dit" sound, the Morse code signal for N, which represented *no*. Dialing one-four produced the "dit-dah" code signal for A, signifying *yes*.

The Speech Indicator had many inherent problems. For example, it often worked all right for local calls, but out-of-town calls were frequently accompanied by other tones produced at the telephone plant. Weitbrecht also learned that deaf people were unable to use the device when someone accidentally moved the telephone dial after contact was initiated. There was no way to explain any erratic movement of the needle, such as that from accidental noise or motion on the speaker's end of the phone line.

Tales of misunderstanding were common with these awkward devices. National Association of the Deaf President Robert G. Sanderson, for instance, used a Speech Indicator to reserve a flight from Washington, D.C., to his home in Utah. After about ten minutes of making arrangements, he asked if he could hang up. The response was "no," or one needle movement. He then asked if he had missed something, and the answer was "yes, yes," or two needle movements. Sanderson then repeated the date, time, flight number, and airline, but he continued to get one needle movement when he was ready to hang up. After nearly one hour of effort, he finally found a hearing person to come to the telephone—only to have the agent ask him his name.

Another friend of Weitbrecht's used the code with hums. A single hum meant *no*, two hums *yes, yes*, and three hums *please repeat."* When ordering a can of paint from a local hardware store, he asked the salesperson to hum once for each dollar so he would know the price of the paint. Nine dollars meant he would hum nine times. But the salesperson on the other end of the line made a mistake in the number of hums and could not start the count over again. After trying again, both ends of the line gave up in frustration.

Weitbrecht was similarly disappointed when he tested a device called SCOTAC, which translated speech sounds to tactile vibrations. Using this device meant having to train himself to distinguish sounds through the sense of touch. "It is not surprising that our reaction was as stated—indifferent. . . . We adults are quite busy in our varied endeavors and feel that we do not have sufficient time to try to learn the tactile vocabulary."[8]

When Weitbrecht looked to the mammoth American Telephone and Telegraph (AT&T) to see what it had to offer, he learned that despite a long history of research in telephony, the company still provided no adequate telephone services to profoundly deaf people.[9] AT&T's "Tactaphone" revealed problems similar to those he found with the speech indicating signalers. Tactaphone was a device that vibrated when someone spoke. While the device let a deaf person know that another party was speaking, it did not allow people on the other end of the line to say much of anything. (The Speech Indicator had a similar problem.) Noise and accidental utterances caused confusion, and the most basic conversation was often reduced to a game of twenty questions.

AT&T's teletypewriter exchange service (TWX) was also a poor choice for the network the three deaf Californians envisioned. TWX used teletypewriters, but most subscribers owned businesses and used the service for data transmission. The service was not only too expensive, it was restricted to subscribers and could not be used to telephone other people.

By 1964, there were more than 85 million telephones in the United States and Canada, but no more than a fraction of one percent of the nation's deaf people used telephones independently on a regular basis, even with the gadgets available. Therefore, it was not surprising that Weitbrecht judged the various alternatives to the TTY impractical. The existing technologies all had significant disadvantages in comparison to the acoustic coupler approach he was pursuing. He knew that the TTY was a huge stone to roll up the mountain, but it showed great promise of ending the ordeal faced by deaf persons. Besides, he was a mountain climber and the challenge was one he was willing to accept.

With renewed energy, Weitbrecht continued to explore the options available to deaf people. After completing an analysis of alternative telephone devices, he concluded that there were no competing technologies to worry about. One more avenue, however, was suggested to the partnership and then ultimately rejected. Marsters and Saks thought the federal government might be interested in supporting research for a system based on a specially designed modem. Marsters contacted several government agencies about a possible grant, but his description of Weitbrecht's teletypewriter-telephone for deaf people was met with disbelieving eyes. He was advised to pursue speech-to-text technology instead. He knew that this was technically less feasible and culturally more restrictive than the TTY, requiring deaf users to have easily comprehensible speech.

Marsters also knew that throughout history telecommunications technology had had a collective negative impact on deaf persons. In the early 1920s, radio introduced a new form of entertainment that deaf people were unable to enjoy. A few years later, in 1926, Bell System scientists H. M. Stoller and A. S. Pfannstiehl designed the first machine to synchronize sound in motion picture films. The advent of "talkies" meant the elimination of subtitles from the silent movies. Gone was a great pleasure for deaf people, who had gathered at local theaters for years. Television, too, had remained inaccessible to deaf viewers for decades. These "advances" in technology were for hearing people. As society changed its long-distance communication patterns, deaf people became increasingly isolated.[10]

Weitbrecht, Marsters, and Saks were convinced that breaking out of this isolation by making daily use of the telephone available for deaf people should be their priority. They would not let government disinterest turn them away from their pursuit of a visual form of long-distance communication. The challenge of developing a reliable product was left to Weitbrecht, and he understood from the start how important it was to have a reliable design.

3

SOMETHING OLD, SOMETHING NEW

The marriage of the older TTY to the new acoustic coupler was the crux of Weitbrecht's design for a visual telephone device. He planned to develop a coupler that would produce a different audio signal as each key on the teletypewriter was pressed. Typing a single letter on the keyboard would "modulate" the signal to be transmitted by adding information to it. A similar transmitter-receiver unit on the other end of the telephone line would "demodulate" the signal, converting the information to an electric pulse that would print the same letter on the receiver's TTY. The modulating-demodulating (or "modem") device would have to be built and sold to each deaf person wishing to have a telephone.

Weitbrecht faced several obstacles as he set out to develop the telephone modem. First, the telephone modem could not be connected directly to telephone company equipment. The telephone companies were strict about "foreign attachments." They were concerned that when a customer connected another device directly to the telephone lines, there might be electrical interference with the company's signals. AT&T's restrictions on direct connections frustrated Weitbrecht's attempt to find solutions. He knew that a direct connection to the phone line would reduce garble in the TTY messages. But anyone who attempted a direct connection ran the risk of having telephone services stopped. In an attempt to satisfy AT&T, Weitbrecht spent years conducting experiments with a modem that avoided a direct connection.

Second, it was important that modem use not disable the telephone system in any way. Therefore, Weitbrecht designed a wooden cradle box on which to place the telephone handset. The cradle contained a miniature loudspeaker and an induction coil. The loudspeaker sent

30

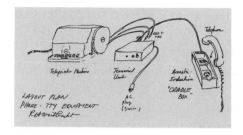

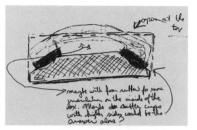

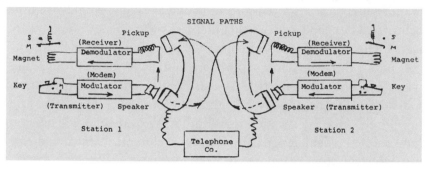

(Top left) Robert H. Weitbrecht's design for the first TTY for deaf people included an acoustic coupler that converted audio tones received from the telephone through a microphone in the *(top right)* cradle to electric signals and produced typed messages on the teleprinter. *(Bottom)* His schematic diagram for signal paths between two TTY stations shows how the modulator-demodulator (modem) would work. Courtesy of James C. Marsters.

tones into the microphone (mouthpiece) of the telephone handset. The induction coil received a message by picking up the signal through the magnetic field generated by the telephone earpiece. With this use of a cradle in his design, Weitbrecht bypassed AT&T's objection to a direct connection to their telephone lines, thus ensuring that the modem would not disable the phone.

Third, he had to keep the modem tone loudness to a minimum because AT&T had established restrictions on the volume of the audible tones that could be generated by modems. This obstacle presented a special problem. The lower the volume of the modem tone, the more garbled the message could become. To make matters more complicated, the volume levels established by AT&T were very hard for Weitbrecht to work with.

Fourth, Weitbrecht had to find a way to signal the deaf caller that the telephone at the receiving end was busy, ringing, or being answered. This was easily solved with a small indicator lamp that lit with the dial tone. Busy and ringing signals were distinguished by the length of time the lamp remained lit.

The scarcity of used TTYs presented a different kind of problem. Based on his amateur radio experience, Weitbrecht knew that telephone companies destroyed surplus teleprinters to prevent them from being used by people who would otherwise subscribe to company services. Yet these discarded machines could provide deaf people with low-cost phone devices. Marsters purchased a new Model 32 ASR (Automatic Send-Receive) Teletype to assist Weitbrecht in the early trials. Marsters' willingness to pay the substantial sum of $600 in 1964 for a new trouble-free TTY was critical. With it, Weitbrecht could concentrate on the modem design and worry less about mechanical and electrical teletypewriter malfunctions.

Still another problem was that the older TTYs spoke a language of their own: Baudot. This was a code developed for mechanized telegraphy in the nineteenth century by an officer in the French telegraph service. It is called a "five-level" code for transmitting and receiving messages because five "bits" of information are needed for each letter, figure, or other character being transmitted. Newer and faster "eight-level" machines use the American Standard Code for Information Interchange (ASCII), which has an extended set of characters.[1]

Using Baudot TTYs in the late 1960s was much like using a slide rule today instead of a calculator or a computer. Although it was functional, the five-level TTY was considered an obsolete device by the corporate world. Still, it would bring immediate telephone access for deaf people who had waited far too long already.

By May 1964 Weitbrecht had developed his first two modems to a point where he was ready to test them. Calling from Redwood City, California, he made the historic first long-distance TTY call on a regular telephone line to Marsters in Pasadena. It was the first time Marsters had used a TTY, and he was especially thrilled to experience the independence. But electrical "echoes" in the telephone line made

Weitbrecht's first modem, built in 1964. Courtesy of James Haynes, W6JVE.

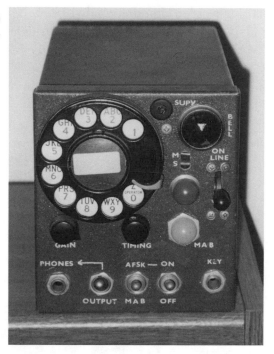

it hard for Weitbrecht to read the words Marsters was typing. As Weitbrecht struggled to keep the telephone message ungarbled, he felt more and more determined. He frequently asked Marsters if he was reading the typed words being sent out through the primitive TTY modem. In Pasadena, Marsters also made various adjustments on his TTY as he sent and received the test messages. Their final messages came across clearly:

ARE YOU PRINTING ME NOW? LET'S QUIT FOR NOW AND GLOAT OVER THE SUCCESS. GOOD NIGHT, JIM. GOOD NIGHT, BOB.[2]

Weitbrecht was confident that the TTY modem would work. There was no mystical process of discovery—only a tremendous amount of experimentation ahead to perfect the system. Above his desk hung a harshly worded sign that guided his work: "The world is full of edu-cated derelicts. Persistence and determination are omnipotent."

After the first call, Weitbrecht worked earnestly on the echo prob-lem. Modems designed for other purposes used a variety of signaling frequencies and patterns. To Weitbrecht, this mixture of signals was a "Tower of Babel among the telephone lines." In long distance signals, this noise can be tolerated by the human ear, but it hopelessly confuses electronic devices that depend on sensitive differences in tones. Elimi-nating the garbled messages caused by random disturbances along the telephone line would require a special component in his modem. He experimented for months and eventually solved the problem with a variable echo suppressor, which included a limiter and a filter to re-duce unwanted signal noise, especially in the spaces between charac-ters. Weitbrecht applied for a patent for his variable echo suppressor,

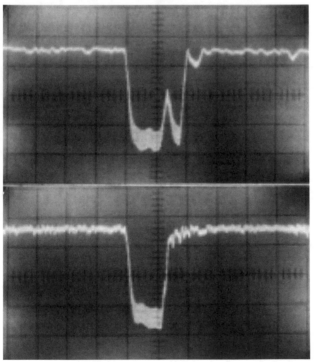

In the 1960s, telephone circuits exhibited a phenome-non of reflection, or electrical "echoes" that garbled the TTY signals. Weitbrecht developed an "echo suppressor," and it was the basis for his patent. The top graph shows the signal be-fore echo suppression, the bottom graph shows the signal after echo suppression. Courtesy of James C. Marsters.

marking the birth of the acoustic telephone coupler and ushering in the world of telecommunications for deaf persons.[3]

Marsters was pleased with Weitbrecht's progress over the first few months of 1964 and saw this as an auspicious time for gaining telephone access for deaf people. Several organizations serving deaf people were being restructured to better address a variety of problems. The National Association of the Deaf had moved its home offices to the Washington, D.C., area to enhance its lobbying efforts with the federal government. The Professional Rehabilitation Workers with the Adult Deaf was also formed when rehabilitation professionals realized that the National Rehabilitation Association, which served all people with disabilities, was not efficiently meeting the needs of deaf persons. Marsters hoped that all these changes would go hand in hand with access to the telephone, serving to improve deaf people's lives.

The year 1964 also marked President Lyndon Baines Johnson's signing of the Civil Rights Act. A public demonstration of the TTY by Marsters, Weitbrecht, and Saks at this time might have linked the cause of civil rights to deaf people's needs. But these three deaf men were not thinking in those terms, and at that point, none desired to make a career of the development of phone devices for deaf people. Marsters had no plans to leave his profession as an orthodontist. Saks had an investment business and, like Marsters, was financially comfortable. Weitbrecht had made a career in academic research laboratories, first at Yerkes Observatory in Wisconsin and then at the Stanford Research Institute. None had experience in industrial research, corporate laboratories, or in receiving government grants. In 1964, all three men were just interested in fulfilling a need. It was thus an impromptu decision to find two TTYs and to demonstrate the telephone for deaf people at the Alexander Graham Bell Association for the Deaf convention in Salt Lake City, Utah, that July.

While preparing for the A. G. Bell Association convention, Weitbrecht read about AT&T's Picturephone, which was being shown at the World's Fair in New York City. He then traveled to New York and saw a demonstration. Weitbrecht understood the excitement generated

(*Left to right*): The Teletype Corporation Model 28 KSR Page Printer, which was used by the military, had the conventional three-bank configuration with numerals, punctuation marks, and special symbols available in the upper case positions; the Teletype Corporation Model 26 Page Printer; Western Union Model 103 TTY. These early model TTYs often were reconditioned for use by deaf people in the 1960s and 1970s. From the collection of I. Lee Brody, courtesy of NY-NJ Phone-TTY, Inc. Photographs by George Potanovic, Jr./Sun Studios.

by the Picturephone—it would allow face-to-face communication, including signing or speechreading—but he knew that regular and practical use of Picturephones was still a way off. More than three decades earlier AT&T had experimented with similar devices that had never reached the market.[4] Weitbrecht had clippings from the 1929, 1954, and 1958 demonstrations of the Bell System video telephone technology that had at first excited—but had ultimately disappointed—deaf people. He stuffed the most recent news release into a file folder. To him, the announcement by AT&T would only create a new set of false hopes. The Picturephone was not only expensive, but its use was restricted to only three locations in the country. "I was impressed by the ease with which I could converse with total strangers over this device," Weitbrecht reported. "However as far as equipment is concerned, Picturephone will not be in widespread use until some years hence—and then the service will probably be quite costly."[5]

The National Association of the Deaf (NAD) convention in Washington, D.C., in 1964 also included an exhibit on telecommunications accessibility, highlighting the strides made in recent years toward helping deaf people "achieve a measure of independence in telephone and other types of instantaneous communication."[6] The Electrowriter and the Bell System Picturephone, Tactaphone, and a direct-connect teletypewriter set up over a teletypewriter exchange (TWX) line were displayed. Martin L. A. Sternberg, a deaf scholar attending the convention, had used TTYs to transmit messages for the Home Service of the American Red Cross during World War II. He immediately recognized the potential of the TTY as the solution to deaf people's long-distance communication needs. "One by one, I brought my Deaf friends in to see this intriguing equipment."[7] Every time, they asked how they might use it.

While at the convention, Sternberg also tried the Picturephone. His open letter to AT&T was printed in the March 1965 issue of *The Deaf American*, the magazine published by the NAD: "With the advent of the Picturephone, something we have been dreaming about for ages, this barrier is about to be broken. We owe you and the other telephone people a warm 'Thank You.'" AT&T's Board chairman Frederick R. Kappel responded that Sternberg's letter was "one of the best rewards my job ever brought me." But he reminded Sternberg that AT&T was still in the early stages of this undertaking; costs were high; and the "full range of public interest in the service is not yet known, and many problems remain to be solved." He assured Sternberg that the value of Picturephone service to deaf people "will always be one of the important considerations that we shall have in mind."[8] Still, even with this cautionary note from Kappel, AT&T boldly promoted the Picturephone in its news releases as a promising solution for deaf people: "To the deaf, who may never have known the convenience of telephone service, it meant much more. In addition to regular voice transmission, Picturephone service transmits the person's image enabling deaf and voiceless callers to converse by sign language."[9] Despite AT&T's publicity and Sternberg's optimism, the more important phone demonstration would take place in Salt Lake City.

To hearing people unfamiliar with the ninety-year plight of the deaf world, a telephone call between two hotel rooms in 1964 seems like an ordinary accomplishment. But to fifty-four-year-old deaf chemist and statistician H. Latham Breunig, it was something entirely different: It was the first independent telephone call of his life. "There, between two rooms in the Hotel Utah," Breunig later wrote, "these deaf people were able for the first time to make unassisted telephone calls over the regular voice grade telephone network by means of a teletypewriter (TTY) in each room."[10]

Like his close friends Weitbrecht, Marsters, and Saks, Breunig had become successful in a professional field unrelated to deafness. A researcher, he used theoretical and applied statistics to analyze chemical procedures for his employer, Eli Lilly and Company in Indianapolis. In 1964, he also was serving on the board of directors for the A. G. Bell Association, and for him this new experience was very exciting. The demonstration call at the association's convention marked the start of a movement toward a new "community"—a spontaneous community

The Bell System Picturephone excited many deaf people in the 1960s, though the technology was not ready for widespread use. The video telephone allowed two people to see each other as they conversed. At a demonstration in the Prudential Building in Chicago in June 1964, student Howard Mann learns about the Picturephone from John D. deButts, president of the Illinois Bell Telephone Company, before making the first call. Property of AT&T Archives. Reprinted with permission of AT&T.

unlike any ever begun in history. Although not yet understood by Breunig, this community for deaf people would be developed *to create* a telephone, and he would play a significant role in it.

As important as Breunig and his other friends were, however, Marsters realized that unity and strength among deaf people and the organizations serving them would be crucial to the phone project's success. Thus he had spent the weeks preceding the A. G. Bell Association convention contacting people about attending the demonstration. He knew that he confronted an uphill struggle in achieving the cooperation he sought, for the National Association of the Deaf and the A. G. Bell Association and their respective memberships had a long history of philosophical differences about how deaf people should communicate. The National Association of the Deaf attracted members who valued sign language as a primary means of communication. The A. G. Bell Association, on the other hand, advocated the use of spoken communication, speechreading, and residual hearing.

Historically, hearing people dominated the A. G. Bell Association, but by 1964 a small group of deaf people had established the Oral Deaf Adults Section. This group formed a communications committee to investigate ways to use the telephone. Weitbrecht, Marsters, Saks, and Breunig were active members, but they needed an alliance with the signing deaf community to help promote what they believed would be a technological breakthrough. Among the influential deaf people they invited to the first TTY demonstration in Utah were Robert G. Sanderson, president of the National Association of the Deaf, and Boyce R. Williams, who represented the Vocational Rehabilitation Administration in the United States Department of Health, Education and Welfare. Williams had worked for two decades to promote employment opportunities for deaf people. More than any other American, he was responsible for better training and improved mobility and job placement for deaf workers.

The National Association of the Deaf was an especially important ally. Established in 1880, it was the oldest self-advocacy organization for a group of people with disabilities in the United States. In 1964, the organization was in the process of forming state chapters so that it

Robert G. Sanderson *(2ⁿᵈ from left),* president of the National Association of the Deaf, types a telephone message with *(left to right)* Robert H. Weitbrecht, James C. Marsters, and Andrew Saks looking on. Courtesy of National Association of the Deaf.

could deal more effectively with civil rights issues for deaf people. Like the A. G. Bell Association, which met just one week later, the NAD had begun to make the issue of telecommunications access more visible to society at large.

Inviting the NAD president and other representatives to the Salt Lake City demonstration, while the modem was still in its design stage, indicated foresight and acumen. Sanderson was an influential individual. He not only had the respect of the signing American deaf community, but he also was coordinator of services for the deaf in the Utah State Board of Education. He was clearly interested in pursuing the TTY further. During his first TTY conversation, he typed to Breunig,

THE TELETYPE SYSTEM YOU FELLOWS HAVE WORKED OUT . . . IS QUITE PROMISING. HOPE THAT WE CAN DEVELOP IT TO THE POINT WHERE ALL DEAF PEOPLE WILL BE ABLE TO AFFORD IT.[11]

He followed this by publishing a summary of his first TTY call in *The Deaf American* and praising the men, all A. G. Bell Association members, who had designed it: "It was a most remarkable experience, and in the years to come I am sure that many deaf people will have reason to appreciate the work of these men and their associates who developed and tested the device and carried it through to success."[12]

While the TTY sparked interest among those who witnessed it in Salt Lake City and established valuable links between two sometimes antagonistic deaf groups, the demonstration was deliberately low-key. The three Californians were nowhere near ready to distribute modems and TTYs in quantity. Indeed, two TTYs had to be borrowed from the Mountain Bell Telephone Company for the Salt Lake City demonstration. After returning to California, Marsters wrote to Sertoma, a civic service organization with a history of working with deaf people. Marsters hoped Sertoma might consider collecting and storing older TTYs as they were phased out by Western Union and telephone companies, but nothing came of this effort.

There also was some resistance from within the Alexander Graham Bell Association itself toward the TTY and modem demonstrated at Salt Lake City. "Certain people," Weitbrecht explained, "object in the way that it does not measure to [the association's] ideals of promotion of lipreading, speechreading and utilization of the residual hearing of the deaf." He elegantly argued against such resistance: "We have here an effective communication method over the telephone circuit between any body concerned, whether normal hearing or deaf. It is particularly noted that any such system enables us deaf people to be completely independent of 'ancillary personnel' [hearing people], and this is to be desired." To this he followed with a personal note, "I, for instance, live alone and have no hearing help to cooperate in handling my telephone calls."[13]

By December 1964, Weitbrecht and Marsters were still the only owners of TTY stations. The seeds for a network had been sown at the Salt Lake City convention, but on very rocky land. The bureaucratic AT&T continued its systematic destruction of the older machines.

Radio hams, however, were also salvaging TTY parts. Wally Buckley, whose call sign was W6GGC (the last three letters standing for "Golden Garbage Can") ran a scrapyard at Third and Evans Street in San Francisco and sold TTYs to other hams. Unfortunately, the telephone companies stopped him by sending the TTYs to San Leandro, where they were crushed for scrap metal. The move was a blow to the hams, and Weitbrecht wondered how difficult it would be to convince AT&T that deaf people needed the machines as their sole means for telephone access, not for a hobby.

In spite of this setback, Weitbrecht continued to move forward with his plan to establish a phone network for deaf people. In January 1965, he finally found a TTY for Saks. Weitbrecht was pleased to have the assistance of the deaf engineer. Just before Saks received his machine, Weitbrecht phoned Marsters: **NO DOUBT ANDY IS DELIGHTED AND IN SUSPENSION WITH ANTICIPATION, EH? HE IS ONE WITH A MECHANICAL BENT.** Marsters felt the same way about Saks: **YES THAT FELLOW IS A MECHANICAL GENIUS, NO FOOLING, I AGREE.**[14] The deaf phone network had now grown to three.

Saks's father, William Saks, who died when Andrew was a teenager, would have been proud to see his son pursue this technology. In 1922, when Andrew was only four years old, his father had written to Alexander Graham Bell seeking advice on educating the boy: "No one now living would be as qualified as you are to help me in this matter."[15] Bell responded, "I am always glad to be of assistance in questions of this kind,"[16] and provided some suggestions. Now, Andrew Saks was in a position to help Weitbrecht and Marsters bring Bell's telephone to the world of deaf people.

Yet it was still very slow going, and the process seemed constantly on the verge of collapse. Weitbrecht, Marsters, and Saks could only devote their spare time to this effort, but they had to find ways to give the network momentum—or the whole notion of a telephone for the deaf might be lost. They decided that further publicity might help build the network and make the purchase of a TTY more meaningful to those who questioned the cost. The three Californians sent short articles to periodicals published by organizations serving deaf people, encouraged reporters to publish stories in local newspapers, and tried other, more novel approaches. For example, the first week of February 1965, Saks gathered his family and some friends together for a demonstration of his TTY at the home of John Buchans, the father of deaf twin boys. Buchans was the vice-president of AMPEX Corporation in Redwood City, a leading company in the production of tape-recording and other technical equipment. Saks saw this "home demonstration" as a more effective strategy for spreading the news about the modem than the newspaper articles.

The "deaf grapevine" has always been a remarkable way to spread information, even in the days before telephone access. Many deaf schools had their own magazines going back to the mid-nineteenth century, collectively known as the "Little Paper Family." When something important needed to get around, it happened quickly through these magazines, telephone calls made with the assistance of hearing family members, postcards, and face-to-face communication in deaf clubs. Saks hoped that would be the case for the TTY.

The enthusiasm at the home demonstration was contagious. Saks wrote to Marsters, "Sunday night at the Buchans was a huge success."[17] Saks's wife Jean and son Bill were "amazed at the potential of the teletype equipment." Everyone took turns typing to Weitbrecht in Redwood City. Jean Saks was so fascinated with the telephone that she immediately asked her husband to search for three more TTYs—one for their own home, one for her mother, and one for his mother. Born deaf, she had previously experienced the benefits of other technologies. By the time she was thirteen years old, she was lugging around a hearing aid as big as an overnight suitcase. The bulky TTY meant more to her, though. After so many years of waiting, she saw the day coming when she would be able to communicate on the telephone without her daughter Andrea's help.

Ever since the voice telephone was invented by Alexander Graham Bell, deaf parents had had to depend on their hearing children to make calls. Jean and Andrew Saks had been dependent on Andrea to make telephone calls for them since she was three years old. At work, her father employed a secretary who listened on a telephone and silently mouthed the words as he read her lips. He would then respond by speaking into a second telephone handset. At home, however, Andrea served in this helping role. Andrew Saks resented having to depend on the intellect of a child during his business calls, however, and he and Jean felt demoralized by their dependence.

Following the demonstration at the Buchans, Saks asked Marsters to tell Weitbrecht he would be happy to help assemble the extra telephone modem circuits: "I am an old hand with the soldering iron."[18] But this was not the time for acquiring extra TTYs for family members. The supply of TTYs was very low. The radio hams were grabbing any that became available, often not aware of the search by the small group of deaf people. Weitbrecht convinced Saks to wait before acquiring additional TTYs. It was critical to focus on installing TTYs in locations that would help get the word out and encourage others to get involved.

Because Weitbrecht was anxious to gain support for the telecommunications movement worldwide, he decided to make the first cross-country TTY call by the second week of February 1965. This experiment by deaf people came fifty years after Alexander Graham Bell made the first voice telephone call over a transcontinental line from New York to his assistant Thomas Watson in San Francisco. Weitbrecht also chose a California–New York circuit for his TTY test. Using a hearing person to contact the operator, he routed the call from San Francisco to New York and back to Marsters in Pasadena. After finding the quality of the signal excellent, Weitbrecht was enthusiastic:

> I KNOW THAT THIS PHONE . . . TELETYPE IS A BLESSING TO BOTH OF US. . . . I NEVER REAL-IZED HOW USEFUL A TELEPHONE COULD BE EVEN TO DEAF PERSONS WHO CANNOT RECEIVE VOICE ON IT. SURE VERY HANDY.[19]

After that call, Weitbrecht traveled to several cities in the East and called Marsters to test the TTY under various conditions.

There were no announcements of these calls in the newspapers. Few people even knew they were being made. They were quiet experiments that helped Weitbrecht determine that the telephone modem signals did not weaken significantly or create more problems over long distances. He was now confident that the new modem could be used over regular telephone lines. The tests assured him that it was feasible to install TTYs around the nation. But would corporate America help supply them with machines? For each modem they built, they needed a teletypewriter. Without both a TTY and a modem, there would be no "telephone for the deaf."

4

THE CORPORATE WINDMILL

AT&T was not likely to be enthusiastic about the telephone for deaf people proposed by Weitbrecht, Marsters, and Saks. Despite Weitbrecht's efforts to develop an acoustic coupler that would not conflict with AT&T's rules, the phone company remained concerned about "foreign attachments" (non-Bell equipment) introducing interference into the telephone circuits, but the Baudot versus ASCII code issue was even more problematic in early 1965.

Weitbrecht, Marsters, and Saks were planning extensive use of the five-level Baudot code at a time when the telephone company was phasing it out. A long battle over codes among AT&T and other companies had just ended. AT&T saw eight-level ASCII as the wave of the future. Baudot machines were slow, transmitting at a rate of sixty words per minute compared to the one hundred words per minute rate of ASCII machines. For deaf people's needs, the speed of the machine did not matter, but each TTY had to run on the same speed for a conversation to take place over the phone line. For AT&T, it was important that as much momentum as possible be harnessed during this early stage of the transition to ASCII. Thus, there would be little interest in assisting deaf people in the development of an antiquated technology.

Even the promise of increased revenue from long-distance calls did not outweigh the importance of the ASCII issue. If deaf people would agree to use ASCII equipment with Bell system "Data-Phone" services, AT&T *might* have been more interested. But this meant that deaf people would not only have to compete with major industries for the expensive new ASCII machines, they would also have to subscribe to costly services. This was not a solution to the problem of having available and affordable telephone devices.

Even a new Baudot TTY, costing $600, was far too expensive for most deaf people, whose annual income averaged about $3,000 in 1965. There was also a backlog of many months for purchasing these machines, which were still in demand in industry. On top of the purchase price would be the cost of the Weitbrecht modem (not yet determined) and monthly phone bills and other service charges. The overall cost for access to the common telephone for deaf people was still a major roadblock.

To ease this situation, Marsters attempted to increase the supply of used TTYs. But when Marsters approached Pacific Telephone and Telegraph to solicit their cooperation, he learned that the company was concerned about violating the Fair Trade Act by competing with Teletype Corporation, a subsidiary of AT&T. Therefore, a pact had been signed stipulating that TTYs would not be resold. Marsters wrote to Saks: "I must advise you that it is going to be more difficult to secure old TTYs, as the Asst. Vice-President of the [Pacific Telephone and Telegraph Company] in San Francisco sent out an order as of Feb.1st not to sell any more TTY equipment to either individuals or groups."[1] Marsters nevertheless argued with Pacific Telephone and Telegraph representatives that this pact applied only to TTY equipment made during World War II, about two thousand Model 19s. There were few of these machines left; nevertheless, he lost the argument. The giant corporation stood steadfastly in his way.

Thus the dilemma in 1965 was not only that deaf people were limited to the Baudot TTYs, but that they also were limited to used Baudot machines, which were often twenty-five years old, difficult to find, and full of worn-out parts. But Weitbrecht had not made the Baudot decision casually. He had decided on Baudot for good reasons and had been careful to check with deaf leaders as he was struggling with this issue. He also had corresponded with engineers at Teletype Corporation, a leading firm in the telecommunications field, to evaluate the availability of both ASCII and Baudot TTYs. He knew that the use of

Baudot was a stopgap measure, but he believed that tens of thousands of affordable Baudot TTYs would become available from AT&T and Western Union as they switched to ASCII machines. He also thought that newspaper alliances, government-military networks, and other groups utilizing teleprinters would soon be giving up their Baudot TTYs. In addition, Weitbrecht was investigating the compatibility between Baudot and ASCII and sought an "electronic handshake," so to speak. He tested his own modem and found it capable of handling ASCII. He hoped that when the time came, Baudot and ASCII TTYs

```
'9446 ?75 8 -. ,95 43:38=8,0 £343.. ARE YOU CONNECTED OK

I AM NOT RECEIVING HERE........GA


SORRY BUT THERE IS NOTHING COMING THROUGH HERE,. D  SUGGEST YOU RING
OFF AND TRY AGAIN LATER. SKSKSK

WRRM ?  UB EIB
    UZUVVVAO

                        K
W MXM                UUZ EAX UA


HELLO THERE ........ I THINK THERE IS SOMETHING WRONG WITH THE LINE
AS YOU ARE MNOT CLEAR OR COMING THROUGH HERE.   SUGGEST A E
CALLING BACK LATER.    BGA

BIBIN   SKSKSKKSKSKSKSK


   E

   T ZCXAWM VO

YOU ARE STILL NOT COMING THROUGH.
CUMNEVSTUAM
ZM              ENMDUBMEUBLA

I DO NOT KNOW WHO YOU ARE . PLEASE CALLL BACK IN AN HOUR OR SO
WHEN THERE WILL BE SOMEONE HERE TO ANSWER THE PN E PHONE.   GA SKSKSK


ZMIUUM EVLIE
            KMNZVLIKPEUZFO
VÆ                          AOVOICQO UA

YOU ARE COMPLETELY GARBLED.    SKSKSKSK


VH
```

The carriage return and the line feed on the early electromechanical TTYs had to be advanced manually. This, coupled with other problems, made messages difficult to decipher.

might communicate with one another, making the transition between the two codes less painful.

Weitbrecht also considered the possible use of his modem with computers. He conducted various experiments and reported satisfactory results. Finally, he foresaw the logical successor to the electromechanical TTY as a "Television Typewriter" and worked on a TV-TTY compatible with ASCII. But with this, too, the cost issue loomed. Immediate telephone access had to be the priority, and the discarded Baudot TTYs were the only recourse.

Disappointed with the lack of a solution for the availability and affordability problems, Weitbrecht temporarily turned to the third goal, portability. He knew it was unlikely that AT&T would be willing to develop a compact machine, but he had to give it a try. In May of 1965 he hauled a TTY in pieces across the country and assembled it at the home of William Bernstein, a deaf advertising businessman who lived on Long Island, in New York. To test the setup, the Bernsteins called James and Alice Marsters in Pasadena. John Tracy, the deaf son of actor Spencer Tracy, and Marsters' roommate at the Wright Oral School, was having dinner with them that evening and witnessed the call. With Bernstein's station, the network had now increased to four TTYs.

The following day, Weitbrecht and Bernstein met with officials of AT&T. It was important to know how the telephone company would respond directly to their questions about the Weitbrecht modem. Weitbrecht explained that all that was necessary for deaf people to have phone access was to plug a TTY into his modem and put the telephone handset in the cradle. While Bell System engineers recognized the problems facing deaf people, the meeting did not go well. They were not interested in Weitbrecht's modem. Later that day, Weitbrecht called Marsters on the TTY from Bernstein's home and noted, **LOOK AT OUR ACCOMPLISHMENTS . . . I CERTAINLY SEE NO PROBLEM ON OUR SYSTEM OVER THIS LONG DISTANCE CALL . . . JUST AS PERFECT AS IF WE WERE ON THE S.F. [San Francisco] AND L.A. CIRCUIT.**[2]

Weitbrecht was disappointed in the lack of enthusiasm by AT&T officials. He summarized the meeting to Marsters:

WE CERTAINLY HAVE A PROBLEM SHOVING THE WHOLE
AT AND T AN INCH WITH A CROWBAR. THE BUILDING
IS SO HEAVY . . . AT LEAST WE HAVE A WIT-
NESS. . . . THESE TELCO PEOPLE WERE VERY
CLOSEMOUTHED AND I WAS AFRAID TO DISCUSS
MORE BECAUSE OF THE ATMOSPHERE.[3]

His skepticism was justified. In early June 1965, he received a letter from AT&T products planning engineer W. Schiavoni, who explained that the Bell System engineers were familiar with the operation of TTYs over the regular telephone network through the use of coupling, but AT&T was not interested in marketing the equipment: "We have carefully considered the possibility of offering portable instruments to hard-of-hearing and deaf customers, which I understand you advocate, but the conclusion we reached every time the matter has been brought up is that the telephone companies should not offer such portable devices that have no electrical connection to the telephone network."[4]

Schiavoni's letter angered Marsters, Weitbrecht, and Saks. To them, AT&T's response echoed Western Union Telegraph Company's failure to seize an excellent opportunity ninety years earlier. When Gardiner Greene Hubbard offered the company the rights to the telephone patent for $100,000, Western Union president William Orton is purported to have asked, "What use could this company make of an electrical toy?"[5] Now, it was AT&T that seemed shortsighted by not taking advantage of the potential opportunity for revenue growth from deaf customers.

The only bright side in AT&T's lack of interest in TTYs was that the company did not formally object to the use of Weitbrecht's modem by deaf people, as long as the audible tones did not interfere with other telephone services. In fact, Schiavoni had expressed interest in finding ways to give deaf people obsolete TTYs. But this was not necessarily good news. AT&T rejected Weitbrecht's proposal for the development of a new, portable, TTY; it accepted his modem only if it was used with their heavy, not portable, used machines.

Weitbrecht hardly finished digesting Schiavoni's letter before he approached Teletype Corporation engineers to ask them about developing a portable machine. Would they be interested in working with him to develop a special TTY? By consent decree, however, Teletype Corporation, as an AT&T subsidiary, could make only those new products authorized by the Bell Telephone Company. Teletype Corporation engineers had their hands tied, even though some of them were Weitbrecht's friends and did not want to disappoint him.

This experience demonstrated that although Weitbrecht was a wizard with electronics and machines, he did not have the political clout to pressure a large corporation to respond to the needs of profoundly deaf people. Therefore, it was critical for Marsters, Weitbrecht, and Saks to strengthen their alliance with the National Association of the Deaf, the deaf community's consumer advocacy organization. Marsters began searching hard for TTYs for the organization's officers, President Robert Sanderson in Utah and Executive Secretary Frederick C. Schreiber in Washington, D.C.

Weitbrecht, temporarily despondent after his failure with AT&T and feeling that he had let down deaf people, soon regained his optimism and began working closely with his ham radio friends to test the modem further. As the following message, typed one evening on the RTTY circuit to Merrill Swan indicates, the search for a telephone for the deaf was taking on greater and greater importance for Weitbrecht.

IT IS A PRIVILEGE TO BE ABLE TO HELP OUT OTHER PEOPLE WITH COMMUNICATION DEVICES . . . THAT IS MY ONLY AIM IN LIFE. . . . I HAVE ENJOYED TELETYPE VERY MUCH AND WANT TO SHARE IT WITH FRIENDS EVERYWHERE POSSIBLE.[6]

While Weitbrecht refined the modem, Marsters and Saks located and reconditioned several TTYs they had salvaged from surplus stores. Saks also set up a shop in his garage to build cradle boxes to connect to Weitbrecht's modems. One of the new modems went to Joseph Slotnick of Los Angeles, a deaf computer programmer with a degree from Harvard

University, who had acquired a teletypewriter. For a long time, Slotnick and Marsters had the only two TTYs in the Los Angeles metro area. They called each other often and chatted for hours, reams of yellow teletype paper piling up behind their machines.

Weitbrecht's contacts from the ham radio network allowed him to recruit several hearing men with skills that would prove useful. Larry Laitinen, an electronics technician at a local TV station, helped him evaluate the modem over various telephone lines. By August 1965, Ray Morrison at Illinois Bell and Ray Smessaert from Teletype Corporation had also volunteered their technical expertise. Their first assignment was to help install a TTY at the Indianapolis home of Latham Breunig, the deaf chemist and statistician. When the TTY was set up, they monitored the test call from Weitbrecht in Redwood City. As Weitbrecht had expected, technical problems had crept in because of the difficulties in transporting the heavy TTY from California to Breunig's home. These kinds of service problems plagued the early years of telecommunications among deaf people. As Weitbrecht wrote to Morrison, "One problem, as we are too painfully aware, is the maintenance angle on Teletype and teleprinter machines. We are hopeful that some arrangement can be made to take care of this point on a local basis, perhaps through the kindness and arrangement with a telephone company on an outside-hours basis."[7] No such kindness was forthcoming, however.

Weitbrecht's technical expertise was essential to expanding telephone access among deaf people, but the man behind the modem had a full-time position at the Stanford Research Institute and could only spend his evenings and weekend hours on this project. Marsters, who originated the idea of the TTY network, also was unable to manage the ambitious project they envisioned. He was busy with a full-time orthodontics practice, his family, and many community activities, serving as president of the United Crusade Dental Fund and a Senior Examiner on the California Board of Dental Examiners. He regularly flew his private airplane to Lone Pine, California, where he provided low-cost dental care to cattle ranchers and other people who could not afford the services. Therefore, Weitbrecht and Marsters really needed Andrew Saks to take an active role.

Saks was a visionary. On August 6, 1965, he had suggested a clever idea to address the fourth goal they had established—that of linking the visual telephone of deaf persons to the voice telephone network of hearing people. In a letter to Marsters, he described his conception of a telephone relay service, explaining that an intermediary service with a "second TTY should have its own phone number so that the person having it can use his own phone to receive . . . incoming phone calls and relay them . . . on the TTY on its own phone line. Likewise for . . . replies which could be handled the same way in the reverse."[8] The relay service idea was revolutionary and far ahead of its time.

It had now been about a year and a half since Marsters first met Weitbrecht in Redwood City. Marsters was anxious to see better progress, but organization was lacking. Weitbrecht was bogged down in technical work and in developing a patent application; no formal relationship bound the three deaf men together. Marsters expressed his concern to Saks:

I WANT BOB TO GET GOING BUT HE IS PROBABLY WORKING ON THE REFINEMENTS. PATENTING INVOLVES A LOT OF MONEY AND I AM SURE BOB IS NOT AWARE OF HOW MUCH IT TAKES TO SEE A PATENT ALL THE WAY THROUGH.[9]

It was time to find a way to organize and coordinate their individual efforts and to get money together for the increasing expenses. Saks and Marsters believed that a business partnership was the answer. Marsters, who served as the primary liaison to organizations serving deaf people and eventually became the politician of the partnership, handled much of the contact in the battles with industry and competitors. He saw their efforts as possibly lucrative if something could be done about the TTY shortage, though this hope was never realized for these three deaf men. He told Saks, **LATER ON IF THE VENTURE BECOMES**

SUCCESSFUL, THEN WE CAN ALWAYS GIVE MONEY TO WORTH-
WHILE CAUSES.[10]

Weitbrecht agreed with the idea of a business, and in 1965 he, Marsters, and Saks established the Robert H. Weitbrecht Company to design, develop, manufacture, and market electronics devices and equipment. Each partner initially contributed $500 to finance the venture. Later, more and more funds were needed to pay for a patent attorney; parts and printed circuit boards for the remainder of the first batch of eighteen modems, which they had begun to assemble in Saks's garage; and ever-increasing development costs of all kinds, but this was not foreseen in 1965.

Jean Saks was an enthusiastic early proponent of the TTY. Here she is seen typing on a Model 15 TTY with *(left)* Alexander Graham Bell's granddaughter Lilian Grosvenor Jones and *(right)* Dr. Helen Lane looking on. Courtesy of Jean M. Saks.

For a while, the small Robert H. Weitbrecht Company operated out of the office owned by Andrew Saks in Menlo Park, California. Then Jean Saks located a large office with a warehouse in Belmont, California, and the business was firmly established. By October 1965, Weitbrecht was satisfied that his refined modem now eliminated echoes coming and going over the telephone line. He recommended that they begin production of additional units. PRINTEX in Mountain View, California, became the first company to manufacture printed modem circuits for the TTY. Assembly of the modem units progressed slowly, however, and as long as the TTYs were difficult to obtain, large-scale manufacturing of the modems was out of the question.

The failure to interest AT&T in manufacturing a special TTY at reasonable cost was a major setback for Weitbrecht. Like the fictional Don Quixote, he seemed to be battling a windmill. Month after month, he was fighting the different battles as they came up. He would not give up. In September 1965 he wrote again to Ray Morrison at Illinois Bell, "Thus far, experience with teletypewriter equipment placed in homes of deaf people shows that it is a very welcome and acceptable mode of communications. I wish to encourage Teletype Corporation to look into the possibility of manufacturing a simplified and inexpensive yet acceptable teleprinter machine for use by private individuals."[11] He hoped that the advent of time-shared computer systems might accelerate the development of such machinery in a form deaf people could use. An idea that turned out to be ahead of its time, he proposed building his modem "into the [TTY] itself . . . I think I could cram [the printed circuit boards] in the nooks and crannies."[12] Again, he failed to interest AT&T.

Weitbrecht was also interested in helping Morrison with a TTY for people both blind and deaf. As early as 1963, Morrison had published an article entitled "A New Braille Teleprinter System" in RTTY magazine.[13] Weitbrecht offered two acoustic couplers for Morrison's experimental work, "I would be most happy to cooperate all I can with you to develop suitable terminal equipment . . . to enable intercommunication between Braille Teletype machines over telephone circuits."[14]

By the end of 1965, it seemed to Weitbrecht, Marsters, and Saks that the telephone industry would never support their ambitions. But they kept going, working long hours, with rewards few and far between. The initial hope that some day their business might take off and that they would reap some financial benefits was quickly evaporating. Instead, they settled in to the arduous task of bringing the telephone to deaf people in whatever manner proved successful. They wondered often if they would succeed, feeling shunted aside by a powerful corporation that lacked their vision.

5

THE FRUSTRATION GROWS

The Robert H. Weitbrecht Company partners were not alone in their disappointment with the telephone companies, of which there were about two thousand in the mid-1960s. Not all were part of the nationwide Bell Telephone System, but none made any effort to help profoundly deaf people access telecommunications. The TTY network had grown to only six stations during the first year, and frustration was mounting in the deaf community.

In October of 1965, Robert G. Sanderson, the president of the National Association of the Deaf, lost patience with the ineffective gadgets AT&T was marketing for deaf people. In an editorial he targeted the most recent device, Sensicall, which was selling for $25, with a monthly service charge of $3, in addition to the regular service fee. Developed by a Garden City Park, Long Island, engineer with the New York Telephone Company, Sensicall was a code device that included a small lamp that flashed when the person on the other end hummed, whistled, or tapped on the handset. Sanderson particularly criticized the service charges: "This sounds like highway robbery to me because it does not cost the telephone company a single dime to have this device in operation . . . I do not see why deaf people should be penalized for trying to make increased use of the telephone."[1]

That same month, Jean Leigh, a member of the Alexander Graham Bell Association for the Deaf, wrote a letter to the editor of *The Volta Review* noting that various devices offered promise for profoundly deaf people, yet no great help appeared forthcoming from the telephone companies. Electrowriters were being used by several deaf people on Long Island, at Public School #47, the junior high school for deaf students in Manhattan, and at the Office of Vocational Rehabilitation, where a deaf counselor was employed. A few were using TTYs through AT&T's expensive TWX service, and Bernstein had a Weitbrecht modem. Each of these devices represented a tremendous personal struggle on the part of individual deaf customers to gain rudimentary telephone access. The lack of interest in the TTY modem was particularly odd in view of its potential to increase telephone company revenues. "Why is the telephone company so modest?" Leigh demanded to know in her open letter.[2]

Weitbrecht, Marsters, and Saks shared Leigh's consternation. As 1965 came to a close, an article in the Bell System's periodical, *Bell Telephone Magazine,* admitted that "At the present time, there is no standard Bell System equipment for use by the totally deaf."[3] Yet more than thirty years earlier the Communications Act of 1934 had promised "to make available, so far as possible to all the people of the United States, a rapid, efficient, Nation-wide, and world-wide wire and radio communications service with adequate facilities at reasonable charges."[4] Furthermore, AT&T claimed that its goal was "universal service." Apparently this universe did not include deaf people.

What puzzled the trio at the Robert H. Weitbrecht Company the most, though, was how AT&T continued to invest money into research on gadgets that would provide access only through the use of codes. All these devices required a great deal of effort to produce any meaningful communication. Since 1963, the company had been actively promoting the use of Morse code devices. In one Bell System news release, AT&T reported, "One of the most ingenious communications services that the Bell System offers the handicapped is a beehive flashing lamp that enables the totally deaf to hold a two-way conversation by phone. It converts dial tone, busy signals and humming sounds into visual signals of varying lengths . . . Two people with a little practice can hold an understandable conversation between themselves by humming Morse or some other mutually understandable code, thereby flashing the lamp."[5]

Ingenious? Deaf consumers did not think so, but the more they heard about the TTY and the Weitbrecht modem, the more interested they were in those devices. By September of 1965, Weitbrecht felt that he was being overwhelmed with requests for assistance in acquiring teletypewriter terminals and modems, and he was getting weary from responding. He wrote to Breunig in Indianapolis: "In a way, I enjoy writing such letters—however I find I have less time to devote to my laboratory activities. This will have to be resolved somehow or other."[6]

Deaf Europeans had also learned about the TTY and became anxious to see it work. In the summer of 1966, Marsters was invited to make a presentation at the University of London. Traveling at his own expense with his wife, Alice, and their three children, he brought a film and slides about the breakthrough in telephone communication in the United States. He discussed technical aspects of the TTY modem with professors and engineers at the University of London and the University of Manchester. He then traveled to Spain, Italy, Austria, and France. Everywhere he generated enthusiasm. Newspapers splashed headlines of the breakthrough: *"Das Telefon der Gehörlosen ist da!", "Des sourds qui téléphonent?", "I sordi potranno telefonare con il sistema della telescrivente", "Teefoon Conversatie Voor Doven."* The editor of France's *Le Messager* described the TTY as "the writing telephone that every one of us would like so much to have."[7]

But Marsters knew that it would be years before deaf people in Europe would experience the pleasure of using a telephone. Bureaucratic governments, complicated telephone networks, and the difficulty in procuring TTYs and manufacturing modems made widespread use of the TTY nearly impossible in other nations. Many countries wanted to develop their own TTY systems, making them exclusive but incompatible with others. This problem of technological nationalism hindered international telecommunications for deaf people for years.

Back home, Weitbrecht, Marsters, and Saks continued to face a nearly unending series of tasks. They contacted telephone companies and surplus stores in search of teleprinters. They spoke to the Kiwanis, the Lions' Club, the Rotary Club, the Telephone Pioneers of America, and many other organizations to try to find support. They independently gave small group TTY demonstrations at schools, churches, and conventions. Weitbrecht's skill was in particular demand. The few TTYs in use included a motley assortment of models requiring him to carefully study the differences in their technical specifications. For their part, Saks and Marsters creatively introduced several beneficial services that would eventually grow with the TTY-telephone network.

One of the most important of these services was the telephone relay, which would give people with TTYs the ability to communicate with people who had only voice equipment. Both Saks in his investment business and Marsters in orthodontics were challenged by the fact that they could not communicate readily by telephone with hear-

James C. Marsters and Robert H. Weitbrecht rebuilt teleprinters from scrap parts. Along with Andrew Saks, they trained others to recondition TTYs. Courtesy of James C. Marsters.

ing people. The invention of the modem was of little help without some kind of bridge between the human voice and the TTY. In the summer of 1966, Saks set up the first TTY answering service with the Tel-Page Company in Redwood City. It was a closed service for a small group of deaf subscribers, including Weitbrecht. Shortly thereafter, Marsters established a similar arrangement with Alert Answering Service in Pasadena.

Unfortunately, these early relay services were not entirely successful. The staffs of both Tel-Page and Alert Answering complained about the bulky and noisy TTYs in their work areas. Costs were a further deterrent. Most answering services for hearing people charged a flat rate, but incoming and outgoing calls beyond established time limits cost extra. Therefore, deaf customers had to pay additional charges for the longer time needed to type messages. They also were charged a monthly rental for floor space for the TTY, plus a special fee for the second telephone required for relay communication. Monthly service bills ran as high as $65 to $100. A few deaf professionals, such as Marsters and Saks, could afford such charges, but most people, hearing or deaf, could not. To keep costs down, a government solution was necessary.

Other problems continued to plague the small company. By 1966, Weitbrecht and Marsters were increasingly worried about competing technologies. In addition to AT&T's gadgets, flashing and vibrating coding devices from independent experimenters refused to go away. Weitbrecht scrupulously checked out each new device that came along. One was "Dialcom" developed at the Speech and Hearing Center at New Mexico State University with funds provided by Captioned Films for the Deaf. With Dialcom, words, phrases, and entire questions were coded into strings of numbers. "Hi. Will you have dinner with me tonight?" required dialing the numbers "11189100111." A simple yes response required "11143111" and the number string "1117151541955211203711" had to be dialed for "Good. See U at 7."

While Dialcom did not present a real threat, Electrowriter was a serious concern. In the fall of 1966, Weitbrecht learned that Gallaudet College (now Gallaudet University) in Washington, D.C., had acquired six Electrowriters. Gallaudet was seen as the most important educational institution for deaf people at that time, and the school's faculty and administrators had close ties with the influential National Association of the Deaf. Weitbrecht therefore pressed his friends from the competing A. G. Bell Association to further strengthen ties with the NAD: "Considering Deaf People as a whole, I am in sympathy with the N.A.D. in this matter [of telecommunications access] and would be most happy to cooperate with that group to the best extent that we are able to obtain and equip teleprinter stations for that group as well as the [A. G. Bell Association]."[8]

The problem of competing technologies quickly worsened, however. Both the Deafness and Communicative Disorders Branch and the Office of Media Services and Captioned Films for the Deaf under the auspices of the Department of Health, Education and Welfare were interested in the Electrowriter. Media Services and Captioned Films for the Deaf provided captioned educational and entertainment films through media distribution centers around the country. These meant much to deaf people, who were unable to enjoy radio, television, or motion pictures. Groups of deaf people congregated, often on a weekly basis, in deaf clubs and private homes to watch films captioned with government funds. The Deafness and Communicative Disorders Branch in the Office of Vocational Rehabilitation (renamed the Rehabilitation Services Administration in 1966) was headed by Boyce R. Williams, one of the American deaf community's most respected leaders. These two offices also served as important contact points for other agencies interested in deafness and deaf people. Marsters, Saks, and Weitbrecht desperately needed their support for the effective diffusion of the TTY.

They realized that the quest for telecommunications access through the TTY would face a major setback if use of the incompatible Electrowriter spread. Marsters got his first taste of competition in this venture while fighting this battle, and he approached it aggressively: "I feel stronger than ever that we should push through our demonstration . . . to the [AT&T] bigwigs in N.Y.C., plus having positive statements from [the National Association of the Deaf], visit with Captioned Films, N.Y. Univ. Center for the Deaf, etc."[9]

A December 1966 demonstration of the TTY at the Washington, D.C., office of the Vocational Rehabilitation Administration was arranged, and Weitbrecht called A. G. Bell Association Executive Director George W. Fellendorf to encourage his support. Fellendorf, a hearing man, was excited about the demonstration. He had acquired one of the Weitbrecht modems, and his deaf daughter had fallen in love with her TTY. She had asked for a dictionary to look up words as she used the telephone. Fellendorf thought use of the TTY might improve her reading and writing skills. He followed up on Weitbrecht's suggestion and contacted the government branches immediately. In a letter to Marsters, Fellendorf thought that "both Captioned Films and V.R.A. are extremely interested in funding the acquisition, storage, and eventual dissemination of the teletype systems so I think we finally have this thing on the road."[10]

For the December 1966 demonstration, Weitbrecht acquired a worn-out TTY from a Western Union plant and rebuilt it. Prior to his departure for Washington, D.C., he called Marsters about the lack of portability: THE MACHINE WEIGHS 76 LBS AND WAS REAL TORTURE ON THE ARMS TO HEFT IT ONTO THE TABLE. SO I HAVE TO TAKE CARE. MY ARMS ARE A BIT SORE.[11] Attending this important demonstration in Boyce Williams's office were Schreiber of the NAD, Joseph Wiedenmayer of the A. G. Bell Association, Walter Marshall of Chesapeake and Potomac Telephone Company, William Usdane, Director of Research in the Vocational Rehabilitation Administration, and Edna Adler, a deaf vocational rehabilitation specialist. The demonstration call to Marsters in California was crucial. It not only helped to gain government support for the TTY, it laid a foundation for bringing telecommunications for deaf people to the workplace.

This demonstration would be a hollow victory, however, if the TTY supply problem could not be addressed successfully. Even as Marsters and Weitbrecht struggled with the challenge presented by the

Electrowriter, they turned to Fellendorf of the Bell Association for assistance with unresponsive AT&T.

Marsters had learned in the fall of 1966 that AT&T planned to trash fifteen hundred obsolete TTYs. He promptly called Fellendorf and asked whether there was anything Fellendorf could do as an officer of the organization that Alexander Graham Bell had founded shortly after the telephone company had started. Fellendorf's first reaction was to call the one person he knew well—a person who had touched the hand that had actually invented the telephone. He contacted Lilian Grosvenor Jones, Bell's granddaughter, and asked her to help.

Jones secured an appointment with the chairman of the Board of Directors of AT&T. She and Fellendorf had a gracious reception at corporate headquarters. Fellendorf left the meeting thinking that the AT&T president and several vice presidents who met with them were supportive of his proposal to salvage the 1500 TTYs headed for the scrapyard. He contacted Marsters to let him know what he thought was good news.

The resolution of the surplus TTY issue, however, was not so simple because two traditions clashed at this meeting. On the one hand, Lilian Grosvenor Jones represented the telephone inventor's lifelong goal—to improve the quality of life for deaf people—and the AT&T executives were sympathetic. On the other hand, releasing TTYs would conflict with another tradition—telephone company monopoly. In the 1950s and 1960s, AT&T had seen bootleg competition grow as small companies began to interconnect customer telephone equipment to the Bell System lines. For example, in 1956, the company was sued by a man named Harry Tuttle, who wished to allow consumers to attach his "Hush-a-Phone," a simple plastic device that fit onto the mouthpiece, making it harder for other people to hear a conversation. Tuttle lost the court case. The FCC ruled that the device violated AT&T's rights and was illegal.

Despite this decision, other devices such as various antique and colored phones were marketed as more and more people viewed their telephones as personal property. AT&T was concerned about the quality of its network and, of course, its revenues. When telephone company

(Top) AT&T destroyed its older machines to prevent reuse. Later, the industry began to discard large numbers of TTYs as faster machines came onto the market. *(Bottom)* Weitbrecht would search in scrapyards for parts to rebuild TTYs. Courtesy of James C. Marsters.

representatives discovered such illegal connections to their lines, they sometimes exercised their right to terminate the customer's service. And Weitbrecht's modem? Although AT&T did not criticize it, its use was dangerously close to a violation. AT&T's position was that any equipment that customers had on their premises, including telephone handsets, were part of the company's "telephone service."

So when no immediate action was forthcoming after the meeting with AT&T, a small group from the A. G. Bell Association began pressuring AT&T officials relentlessly. Fellendorf and Breunig approached AT&T independently in November. Breunig reported that he had no positive encouragement. Marsters, anticipating an invitation to visit AT&T headquarters, wrote to Teletype Corporation in search of a portable TTY: "I expect before long to be called to the headquarters of the AT&T in NYC to demonstrate," he explained. "I wish to have the nicest looking set-up possible."[12] Teletype Corporation engineers loaned him a used portable Model 31 TTY weighing about twenty-five pounds. A few weeks later, Fellendorf met again with AT&T public relations people to discuss further action about teleprinters.

This time, Fellendorf brought with him to AT&T's headquarters a well-known, successful, and politically connected deaf man, Joseph Wiedenmayer. Wiedenmayer had served under President John F. Kennedy as U.S. Consul in Melbourne, Australia, and was awarded the Meritorious Service Certificate from Secretary of State Dean Rusk. He resented the barriers hearing society placed in deaf people's paths. "The educated and experienced hearing impaired person," Wiedenmayer had written, "must be given a chance to compete for interesting top flight jobs and not be barred at the door by arbitrary and precalculated physical restrictions."[13] Wiedenmayer had Usher syndrome, which meant that in addition to his deafness he had limited vision. He was one of the first persons who envisioned a TTY that might have Braille capabilities. But AT&T, embroiled in another connections lawsuit, called the Carterfone Case, still refused to release the TTYs.

By the fall of 1966, eighteen Weitbrecht Company modems were connected to TTYs and operating with little trouble, but the network they provided did not reflect the demographics of the broader deaf community. Modem owners were nearly all well-to-do deaf professionals, including an orthodontist, an engineer, a computer programmer, a chemist, and several business people. Weitbrecht therefore was unable to effectively study consumer use and acceptance of the new technology. He needed a critical mass of TTYs in one location that included deaf blue-collar workers.

In November, he received a letter from Paul L. Taylor that provided promise of such a boost to the network's growth. Taylor was a deaf senior systems engineer at the McDonnell-Douglas Aircraft Company in St. Louis, Missouri. He worked in a reconnaissance laboratory doing analyses on radar, infrared, and photographic devices for jet airplanes. Like Weitbrecht, Taylor and his wife, Sally, had seen the demonstration of AT&T's Picturephone at the 1964 World's Fair in New York City, but their excitement quickly eroded when they were told that the video technology would not be available to the public for many years. In late 1966, they had a private telephone line setup with Sally's parents a few blocks away. The monthly fee for each home was only three dollars, and even though their calls were limited to this local circuit of two rebuilt TTYs, it meant peace of mind. The Taylors finally could telephone someone in case of an emergency.

Taylor learned about the Weitbrecht modem from a friend and wrote to the deaf physicist, optimistically ordering two of them. He then approached T. Alan Hurwitz, also a deaf engineer, about purchasing the second modem for the St. Louis area. But Hurwitz was reluctant to spend more than $200 for the modem. He told Taylor that he did not think it was worth it to pay a telephone installation fee and monthly phone bills in order to communicate with one other family in St. Louis. Hurwitz had grown up with deaf parents and never had a telephone in his home. A next door neighbor had been recruited for assistance when they needed to telephone a doctor or make other important calls. This was a typical arrangement for deaf families.

So Taylor looked elsewhere for someone who might buy the second TTY modem in St. Louis. He told Weitbrecht that he and his wife would like to create a network that would include as many deaf people as possible who desired teletypewriter service. "Since the acquisition of teletypewriters is of no problem to us," he wrote, "it is the installation on the regular telephone wires along with a dialing device that is holding us back. Telephone personnel in the St. Louis area have not been too encouraging on this."[14] Weitbrecht was excited about finding another deaf engineer interested in helping him and intrigued by Taylor's comments. He responded to Taylor immediately, describing the national network of eighteen TTYs and how they operated with couplers over regular telephone lines.

With foresight, Weitbrecht also explained to Taylor how long-distance calls could be made by deaf persons and how their costs were unfair. Telephoning by TTY is slower than by voice. The TTY limited typing speed to sixty words per minute (wpm), and few people could type even this fast. Deaf people thus paid more than hearing customers for transmitting the same information over the telephone. "We think, however," Weitbrecht wrote, "that eventually we will have a special rate for long distance teletypewriter calls—perhaps on the average of one-third less cost, based on rate of information transfer, 60 wpm teletypewriter versus approximately 180 wpm voice-speech."[15]

Weitbrecht was most curious, though, about Taylor's comment on the availability of TTYs in St. Louis, especially since he had been so unsuccessful in finding TTYs in California. He wrote, "May I inquire as to how you obtain such machines? We want to know, as we are sincerely interested in obtaining as many of the teletypewriters as the deaf people may wish to acquire."[16] Taylor responded that Western Union Telegraph Company had agreed to donate discarded TTYs to deaf people in St. Louis, and there was promise of many more from that company through its storage houses around the country. To Weitbrecht, the Western Union TTYs seemed a godsend, but he remained irritated over the lack of progress with AT&T.

AT&T's vacillation and hesitation was infuriating. First they held out a promise and then they dashed it. In late November of 1966, Weitbrecht had thought he might get help from Ray Morrison, an Illinois

Bell engineer, who enthusiastically wrote that a colleague in New York had telephoned him about Weitbrecht's acoustical data set for deaf people. Morrison thought that the potential of the Weitbrecht modem had finally been recognized, and he offered assistance with another demonstration for AT&T. "They will consider my advice seriously," he explained, "so this is a start at AT&T (Bell System) cooperation."[17] As it turned out, it wasn't.

The Weitbrecht Company's December 1966 meeting with AT&T was not successful in getting AT&T's cooperation either. Marsters again showed the AT&T officials how convenient the Model 31 TTY was for deaf people. Wisely, he argued that AT&T had the potential for a new market of deaf telephone callers. But once again, AT&T proposed that deaf customers subscribe to the expensive Bell system Data Phone service instead. Marsters countered that deaf people would only be able to make telephone calls on Data Phone lines to people who were also subscribers, paying the additional fees. He asked the AT&T officials if any of them had a Data Phone line in their offices. No one did. Then he asked them if their mothers had the special Data Phones lines. No again. Finally, he asked them if they had Data Phone lines at home, emphasizing that they were expensive. Once again, the answer was no.

AT&T's resistance apparently was related to their fear of losing their telephone line connection monopoly, which was being challenged in a lawsuit by Carter Electronics. This Dallas, Texas, company was seeking the right to use a two-way radio system called the Carterfone to carry on private communication over telephone lines. The Carterfone case was about choice at the terminal end of telephone lines. It was an effort to end the way AT&T frustrated consumers as well as competition by having the right to disconnect any line that was attached to a non-AT&T product, such as Weitbrecht's modem. Until the case was over, AT&T implied, the release of discarded TTY equipment could not be worked out. In the meantime, the NAD learned that AT&T was not interested in lowering Data Phone rates either. In early 1967, President Sanderson reported, "So far we have received word from AT&T that while the company is cognizant of our problems and will actively work toward achieving our aims, they cannot at this time lower Dataphone rates."[18] Deaf people were put on hold—again.

Even as AT&T dragged its feet, Weitbrecht began thinking about strategies for dealing with TTYs after they became available. He called Marsters: **IF AND WHEN AT&T GOES AHEAD AND GIVES US MACHINES . . . IT LOOKS LIKE A HELL OF A TASK TO PARCEL OUT ALL THE MACHINES AND STORE THEM SOME-WHERE.**[19] Marsters then drafted a planning proposal to make telephone communication more immediately available and more expedient to deaf individuals unable to communicate sufficiently by ordinary or amplified telephones." He suggested that AT&T donate the surplus teleprinters to representatives of the Bell Association, who would pick up and store the TTYs and that a release form would state that the recipients were responsible for maintenance.

Weitbrecht followed up with a paper calling for a nationally coordinated approach to implementing the TTY network and for training people to recondition the machines he hoped AT&T would release. He asked, "Who will install and maintain teleprinter machines in operating condition for use by deaf individuals? . . . "How can such equipment be coupled to telephone lines [and be] available everywhere?" He wondered if the equipment should be restricted to locations installed by telephone companies or whether deaf people could set up their own TTY stations freely once they had telephones. His resolution was prescient: "Probably the best arrangement would be the appointment of regional coordinators, familiar with machine installations, who would be able to travel around, setting up and adjusting machines for a flat fee."[20]

Although Weitbrecht was continuing to plan for a future when more TTYs would be available, by the end of 1966 and the beginning of 1967 the stress and loneliness of his life were affecting him. The ongoing patent struggle and the disappointing AT&T meeting left him disheartened. On Christmas Day 1966, Marsters called Weitbrecht to encourage him by at least relieving his immediate financial worries: **IF YOU ARE WORRIED ABOUT THE COST OF THE PATENT**

THEN PLEASE FORGET ABOUT IT AND LET IT MOVE BY
SENDING WORD TO ATTORNEY. . . . SAKS AND I CAN
WELL HELP CARRY THE BALL. Weitbrecht responded, IT IS
REALLY HEART WARMING TO ME TO KNOW OF THE SUPPORT
THAT YOU FELLOWS ARE GIVING ME . . . AND OUR SYS-
TEM TOO.[21]

Marsters also knew that Weitbrecht was not a careful planner. He
was a talented man, but he lacked the experience and wisdom needed
for effective business or legal dealings. Marsters told Weitbrecht that
he was anxious to get the patent nailed down before something might
happen to influence Weitbrecht's chances of earning royalties and
building a nest egg for retirement: PLEASE BOB LET THE BABY
OF YOURS GET BORN PATENTWISE AND YOULL THANK YOUR
LUCKY STARS.[22]

Weitbrecht was also disheartened by his employer's inability to understand deaf people's desire to have some control over their own lives.
Therefore, his relationship with the Stanford Research Institute grew
increasingly acrimonious. In January 1967, the Institute held a planning meeting to discuss the possibility of sponsoring a symposium to
identify research priorities in the area of communication devices for
deaf people. Remarkably, Weitbrecht was the only deaf person invited
to the meeting. He sensed a threat to the self-empowerment in which
he and his deaf partners so strongly believed. He warned the Institute
not to plan technology for deaf people without their full participation:
"Deaf people, as a rule, are a proud and sensitive group. . . . [and]
should be allowed to evolve their own methods of communication that
aids them to [communicate] in the most expeditious way."[24]

Deaf self-empowerment took hold firmly in February 1967, when
Weitbrecht, Marsters, and Saks decided that they could not wait for
AT&T or any other hearing-owned business to come to the aid of deaf
people. Instead, they would prepare for AT&T's eventual release of
TTYs by establishing their own research and development company,
the Applied Communications Corporation (APCOM). In contrast to
the Robert H. Weitbrecht Company, which they had established in

1965 as a simple partnership, APCOM was incorporated. The company's main focus was on practical telecommunications equipment for use by deaf and deaf-blind people. Marsters was president, Saks was vice president in charge of operations, and Weitbrecht was vice president and director of research and development, handling product design and evaluation. Once again, the partners invested their own money, and they began to market the "Phonetype" acoustic telephone coupler, the name they gave to Weitbrecht's modem, which had just been accepted by the U.S. Patent Office.

At first, both Weitbrecht and Marsters worked only part-time for APCOM, while Saks turned his own finance business into a part-time venture and dedicated himself full-time to APCOM. It was extremely important for Saks's self-worth that APCOM be a success. He came from the high-powered Saks Fifth Avenue family, but even after his father died at an early age, it was a foregone conclusion that, because of his deafness, he would not be allowed a role in the family business name. Managing a business, proving that he could be equal to any hearing man his age, was more important to him than making money. Marsters gained that gratification as an orthodontist. His other close friends in the A. G. Bell Association were successful book editors, psychologists, and businessmen. Saks needed that measure of self-esteem, too.

Saks's role in the day-to-day operations at APCOM was critical. In a sense, he carried on his grandfather's legendary Horatio Alger style — but the riches were intrinsic rewards. He was an organized manager obsessed with the quality of his work. Keeping order in the small but chaotic business, he handled office matters and then often rolled up his sleeves in the evenings to help Weitbrecht recondition TTYs. His secretary, Mona O'Brien, voiced telephone calls for him. Saks would have preferred to use relay services to deal with the many businesses and other individuals who had yet to acquire TTYs, but for the quantity of calls he made daily, the local answering service he had established was inadequate. O'Brien assisted him by repeating the message from the other end while Saks read her lips and spoke for himself.

Saks cherished his old, portable Model 31 TTY, a machine that was used for news and sportscasting. The obsolete machine was not rare, but it represented the dream he shared with Weitbrecht and Marsters of one day owning a compact TTY station a deaf person might carry on a trip. When the Model 31 TTY was stolen from his car one day, he was heartbroken. He had grown attached to it, so he decided to place an advertisement offering $50 for its return. Surprisingly, the thief called him through his secretary and agreed to meet him alone on a dark corner where Saks would buy it back for $75. Against his family's wishes, Andrew Saks got that TTY back.

Saks opened APCOM in a small Menlo Park office room that also functioned as a laboratory for Weitbrecht. Later, the operation moved to a warehouse in Belmont and gained additional space for packing, shipping, and TTY storage. There, Weitbrecht repaired older versions of the modem and conducted research and development work on new products. The J-Omega Company in Mountain View, California, and Hewlett Packard contributed circuit board advice. With this additional help, Weitbrecht was able to improve the Phonetype so that it was reliable for use with hundreds of telephone systems throughout the United States.

As with other new companies, APCOM confronted serious challenges to its viability. It was a small, start-up company that had major expenses for research and development, but the market for its products was unknown. The appropriate consumers, in many cases, were not part of an identifiable group but saw themselves as members of the general public. Furthermore, there was a critical concern about whether deaf consumers would be able to afford any new technology. Like other persons with disabilities, deaf Americans were disproportionately undereducated, unemployed, and underemployed. Thus, it was questionable whether they could afford to purchase special products that were likely to have a high price tag. Low product prices achievable through large-scale production probably could not be applied to a technology for people with a low-incidence disability. The APCOM partners faced these issues long before special-needs manufacturers came onto the market. They realized that they would have to reinvest all of the money received from sales in order to develop new products, purchase laboratory equipment, and pay for the manufacture of parts and for overhead charges. To keep the cost of the new Phonetype modem down, they agreed to serve without salary.

In the spring of 1967, things began to happen. Weitbrecht had not heard from Paul Taylor, the deaf engineer in St. Louis, since November 1966. Taylor's wife, Sally, had given birth to a son, and it was not until May that the new parents could find time to return to the telephone access issue. After Alan Hurwitz turned down the offer to purchase the second Phonetype to start the St. Louis network, Taylor convinced another deaf friend, Richard Meyer, to buy it sight unseen. When their modem arrived, the Taylors followed Weitbrecht's instructions, connected it to a Western Union Model 103 TTY, and tested the setup. Then they called Weitbrecht in California. After the initial experience of making the first independent long-distance call in their lives, hardly an evening passed for months without visitors stopping by to examine the new "telephone for the deaf."

After seeing the TTY work at Taylor's home, Hurwitz realized his mistake. He had not envisioned from Taylor's description how important the TTY could be. But his wife, Vicki, had already decided not to wait. She had gone to the bank, withdrawn her own money, and ordered a Weitbrecht modem and a Western Union Model 103 TTY.

The sale of the modems to Taylor and his friends in St. Louis was an important turning point. Taylor soon became a devoted advocate of telephone access. He formed a small advocacy group called Telephone/Teletype Communicators of St. Louis for the purpose of increasing telephone access on the local level. The local organization was the first of its kind. Taylor's collaboration with Western Union Telegraph Company was also important. The company trained a half dozen deaf people in St. Louis to recondition TTYs, a significant step toward expanding the Missouri TTY community.

Building even a small local TTY network, however, was not an easy task. The few deaf people who had the skills to rebuild the machines faced attitude, economic, and technical challenges. Over the next two months, the Telephone/Teletype Communicators of St. Louis reconditioned and installed only eight TTYs. It took another year to provide 32 families with telephones in St. Louis. After other companies began

Paul and Sally Taylor helped to establish the Telephone/Teletype Communicators of St. Louis, the first local telecommunications advocacy group for deaf people. Also in this photograph is their son David. Courtesy of Sally A. Taylor.

to provide TTYs to the deaf community in St. Louis, the number increased more quickly. As more TTYs arrived from out of town, Western Union employees also offered additional maintenance classes.

Another breakthrough came in June 1967, when the Federal Communications Commission (FCC) ruled that the Carterfone did not impair the safety or quality of the telephone network. The FCC ordered the telephone companies to revise their rules to permit the use of such devices. Now private individuals could connect non-AT&T equipment to the telephone system without fear of having their service terminated.

Demand for TTYs grew rapidly. By the summer of 1967, there were still only about four dozen TTYs rebuilt and in use in the United States, compared with 100 million telephones, one for every two

Americans. Across the country, names began to accumulate on waiting lists for the used AT&T TTYs. Letters came pouring into APCOM as deaf people pleaded for the machines. One woman, for example, wrote that she had tried a TTY in Kansas City and told her husband about it. They were very interested in purchasing a TTY. His job often required that he work overtime, and she never knew when he would be coming home. Every day, she was forced to wait for him as the supper she had prepared grew cold.

In spite of the potential for TTYs, not everyone was enthusiastic about them at first. As was the case with hearing people in the early years of the voice telephone, some deaf people preferred to wait to purchase a telephone after others had done so. Even deaf people in white-collar positions were not won over to the TTY that quickly. Don G. Pettingill, director of counseling services for the deaf at the Callier Hearing and Speech Center in Dallas, Texas, wrote to Marsters in 1967, "I discussed the telephone-teletype system you people have pio-

Deaf people in St. Louis, Missouri, collected old TTYs for use in their telephone network. Courtesy of Sally A. Taylor.

neered. . . . Many times I have explored the possibility of getting such a setup here . . . but it would not be feasible. I have a full-time secretary who acts as my ears on a second phone, and repeats word for word the message just as fast as it comes back to us. This way I can talk to ANYONE without a hitch."[24] He asked Marsters why he would need to change what he was doing. He argued that with the TTY, the other party also must have one, and he just did not talk to enough deaf people to justify the cost. Then, too, he had two hearing boys at home who interpreted for him.

Marsters gently argued that most deaf people were not as fortunate to have full-time secretaries. "It is for this reason the inexpensive telephone-teletype system was developed, since it would tie most profoundly deaf persons in with their friends, jobs, social services as well as doctors."[25] He suggested that perhaps demonstration centers should be set up to encourage deaf people to help themselves.

As Marsters continued to proselytize for the TTY and Saks ran AP-COM, Weitbrecht's professional difficulties only increased. He had asked Pacific Telephone and Telegraph for thirty of its used TTYs stored in San Leandro. Weeks passed, and he still had not heard from them. He became more upset when he learned that the Stanford Research Institute appeared interested in the rights to his modem patent. All along, he had assumed that his independent effort to acquire a patent on Phonetype was acceptable to his supervisors. "I had a chat with Dr. Donald MacQuivey at SRI [about] the RHW Company and SRI hassle over patent papers. Now it is two weeks, and [the two SRI lawyers] have not come up with something." They had promised to put something in writing for him. "I objected to any interference on the part of SRI in our business affairs," Weitbrecht wrote to Marsters. "Apparently this is becoming a difficult point for SRI to resolve."[26]

Weitbrecht's personal life also suffered a major blow when his dog, Blackie, died. Blackie had provided the solitary, socially awkward deaf physicist with devoted companionship. He had traveled all around the country with him. By prompting Blackie to bark, he was also able to test designs of his sound-activated transmitter units. While in Tucson, Arizona, his RTTY friend Walt Nettles gave him a Basenji terrier. The dog was too active for Walt who was older than Weitbrecht. His name was "Uhuru," but Weitbrecht could not pronounce it and changed it to "Bongo." His new dog did not answer to any name, however, and, to Weitbrecht's surprise, he did not bark, either, because the Basenji is a barkless breed. Ironically, Bongo could not help him with his sound-activated equipment testing. So 1967 turned out to be a difficult year for Weitbrecht, even though APCOM and the telecommunications movement was starting to take hold in the deaf community.

6

TELETYPEWRITERS FOR THE DEAF, INCORPORATED

With the Carterfone case finally settled, Marsters recommended that a separate organization be established to locate, recondition, and distribute AT&T's surplus TTYs. On February 20, 1968, AT&T released the first batch—200 TTYs—making distribution an immediate priority. George W. Fellendorf and Joseph Wiedenmayer of the A. G. Bell Association led the talks with AT&T, but the association had been established for the purpose of promoting the use of speech, speech-reading, and residual hearing by people who are deaf or hard of hearing, not to serve as a TTY equipment vendor. So in March, the A. G. Bell Association and the National Association of the Deaf joined forces to create the Teletypewriters for the Deaf Distribution Committee. Their first responsibility was to pick up TTYs from the Western Electric distribution centers of Bell Telephone affiliate companies and store them while plans were being made to rebuild and install the machines in deaf people's homes.

On June 10, 1968, the Teletypewriters for the Deaf Distribution Committee took a major step by incorporating as a not-for-profit corporation called Teletypewriters for the Deaf, Incorporated (TDI). Its purpose was to collect and distribute TTYs under the terms of the agreement with AT&T. Latham Breunig, the deaf chemist and statistician at Eli Lilly in Indianapolis, represented the A. G. Bell Association and was chosen to be president of TDI. Jess M. Smith, a teacher and the editor of *The Deaf American*, represented the National Association of the Deaf and was named vice president.

The first president of TDI, H. Latham Breunig *(background),* was a member of the A. G. Bell Association for the Deaf, and the first vice president, Jess Smith *(foreground),* was a member of the National Association of the Deaf. Courtesy of Telecommunications for the Deaf, Inc.

TDI's formation demonstrated the unifying strength of telecommunications issues. The A. G. Bell Association and the National Association of the Deaf had a long history of opposing philosophies, rooted in nineteenth-century events and attitudes that had persisted far into the twentieth century. The deaf "manualists" in the NAD and the "oralists" in the A. G. Bell Association seemed to have little in common and had no history of working together. The TTY bridged the gap. The machine communicated in printed text rather than either American Sign Language or spoken English.

Smith and Breunig had their own personal experience with this. At the time TDI was established, they were the only TTY owners in Indiana, and they frequently consulted with each other. For these two men to communicate effectively, the TTY was essential because

Breunig preferred speechreading while Smith preferred sign language. In person, they struggled to understand one another.

Working together in TDI, Breunig and Smith made progress. At the outset, they were fortunate to have the support of the Hoosier State Chapter of the Telephone Pioneers of America, a group of retired Bell System employees, who provided office space to TDI at the Indiana Bell Telephone Company building in Indianapolis. The Telephone Pioneers also assisted in distributing TTYs nationwide. TDI served as

James C. Marsters (seated) conducted training sessions with the Telephone Pioneers of America and several volunteers in the basement of his home in Pasadena, California (1968). Courtesy of James C. Marsters.

a nationwide coordinator of "authorized agents," men and women who volunteered to pick up, store, and recondition TTYs.

Breunig's aggressiveness was particularly important to TDI's successful launching. He saw to it that TDI advertised employment opportunities, distributed emergency phone numbers and news, sought other sources of TTYs, answered queries, and maintained a national file of all AT&T-donated TTYs assigned by serial number to deaf people. He developed a national directory of TTY stations from the records-keeping agreement with AT&T and expanded it into a telephone book of TTY numbers, as well as general information on telecommunications and deaf people. Initially distributed in November 1968, this directory represented one of the earliest efforts to identify support services for deaf people. It included a list of 174 TTY owners. Half of these were in California and Missouri, where APCOM and the St. Louis group had implemented local networks. The remaining TTYs were distributed over about twenty states and the District of Columbia. A supplement published three months later increased the number to 288.

It was now four years since Weitbrecht and Marsters had made the first TTY call, and it was ninety-two years since the invention of the voice telephone. In half of the states in the country, there still was not a single deaf person with personal telephone access. Nationwide, only a few deaf people had access to a TTY at work.

One person who was fortunate to have telephone access on the job was Paul Taylor, who used a Phonetype modem to transmit data at the Monsanto Chemical Company in St. Louis. When the modem was freed up, he also used it with a TTY as a regular telephone. A few other businesses also began to recognize the usefulness of the Phonetype modem. One manager at Lockheed Aircraft Corporation in Sunnyvale, California, saw the TTY as a means for establishing understanding between deaf and hearing employees. At the Indianapolis office of Western Union Telegraph Company, another manager described Phonetype as the answer to how to make the network idea feasible.

Weitbrecht considered the symbiotic relationship between TDI and the small research and development company APCOM a common bond critical to the success of the growing volunteer movement.

APCOM created the telephone technology for the network and supported TDI's distribution of the equipment. TDI, in return, expanded the need for APCOM's modem, providing APCOM with revenue. There was a great deal of behind-the-scenes negotiating by Breunig and the APCOM partners with suppliers of the used TTYs. Without their collaboration, it would have been difficult for the local organizations to locate the machines they needed for deaf people, whose names often stayed on waiting lists for months. With its small staff, TDI desperately needed APCOM's help.

There were ironies in this relationship, too. Weitbrecht had come to the TTY through his tinkering with radioteletype, yet Breunig was frustrated that AT&T machines were being given out to hearing people (specifically, radio hams). He penned a note to Taylor in St. Louis complaining "that radio amateurs have been getting Bell System TTYs unbeknownst to AT&T thru Telephone Pioneers!"[1] Taylor's friends shared this feeling of resentment. In the fall of 1968, his group of about twenty people in St. Louis held a banquet to thank Western Union staff for getting them started with access to the telephone and to award Weitbrecht, an RTTY user, a certificate of appreciation. John G. Woodard, a Western Union City Plant supervisor, received a plaque that said, "We at Western Union are happy to see these donated teletypewriters [lose] their identity as 'toys' which was the case when they were being given to the [radio amateurs]."[2]

The donated TTYs were a great help in creating a deaf telephone network, but Weitbrecht continued to pursue the goal of portability. In October of 1968, he wrote to Ray Smessaert, the Project Director in the Products Development Division of Teletype Corporation:

> If possible, could Teletype Corporation introduce two new Product Descriptions covering ready-to-use machines for shipment to (deaf) customers? . . . Any assistance you can give in this direction would be much appreciated. . . . we are anxious to cooperate with you in order to arrive at suitable, ready-to-use equipment.[3]

Smessaert responded with an encouraging comment, "After a review of the units and parts on hand we feel that we have equipment

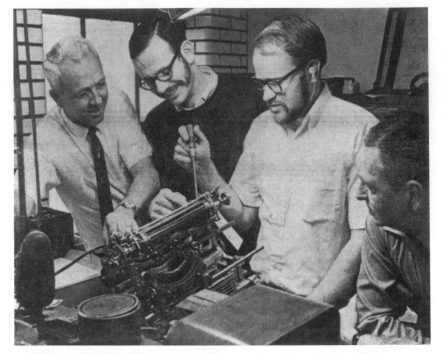

(Left) John G. Woodard, city manager of Western Union in St. Louis, conducted TTY maintenance classes for TDI agents. Each machine required many hours of reconditioning and testing before being installed in a home. Courtesy of Sally A. Taylor.

that will substantially meet the requirements for (deaf) customers, and we feel sure that the minor details can be worked out."[4] Weitbrecht wrote back, "I wonder how much such a special set-up would cost, for direct sale to deaf customers. Would you kindly arrive at a reasonable price?"[5] Three weeks later, Weitbrecht told Marsters that Smessaert was putting a Model 33 into a suitcase. It was 9 inches high, 14.5 inches wide, and 19 inches deep, and weighed about 40 pounds. "I told him we've already done that with a Model 32, but don't appreciate lugging it around."[6] At least there was promise of a "portable" unit on the way. Weitbrecht ordered sixteen new machines. The Model 33 was priced at $788, so only the wealthiest of deaf people could afford it. Still, it was a step forward just to see Teletype Corporation make the modifications on their standard model for use with the deaf network.

The Teletype Corporation Model 32 TTY was lighter (56 pounds) and quieter than the older electro-mechanical models. From the collection of I. Lee Brody, courtesy of NY-NJ Phone-TTY, Inc. Photograph by George Potanovic, Jr./Sun Studios.

Meanwhile, Bell Laboratories continued to ignore the deaf consumer's need for a compact, affordable TTY. At a national forum of the Council of Organizations Serving the Deaf, James Flannagan of Bell Telephone Laboratories summarized the progress on two "experimental telephone aids." One was a simple signaling attachment to the ordinary Touch-Tone telephone, which displayed one letter at a time. The second was a "Visual-Tactile set." The council was made up of representatives from major organizations involved with deaf people. Many of the members were familiar with the promise of the TTY and likely sat through the presentation with a measure of patience and curiosity.

Several years into their project, Weitbrecht, Marsters, and Saks were feeling its effects in numerous ways, some of which could not have been anticipated. APCOM's financial losses forced Saks and Marsters to pitch in more and more money to keep the company going. Because all three partners were bitter about their reliance on hearing people and because they believed the Phonetype modem represented the solution, they were willing to make this sacrifice.

A more positive effect was their growing feeling of independence, which was recognized with some discomfort by the Saks's hearing

John Kubis, a mathematics professor, communicated in American Sign Language with author Harry G. Lang on a Vistaphone at the National Technical Institute for the Deaf (NTID) at Rochester Institute of Technology in the early 1970s. At the time, NTID had the largest number of Vistaphones in the United States. Courtesy of Gannett Rochester Newspapers.

daughter, Andrea, who was accustomed to interpreting phone calls for her parents. As the communication dynamic in her family changed radically, she began to feel a difference in the way her parents treated her. "I was totally lost for a while," she wrote later.[7] She knew that with the expansion of the answering services, which enabled a deaf TTY caller to contact any hearing person with or without a TTY, her parents' independence would be expanded considerably. Her father's initial small-scale experiments with relay services had made this outcome promising—and imminent.

Marsters, too, rejoiced in the freedom to make his own calls. He and his friend Joseph Slotnick continued to expand the local network in the Los Angeles area. They learned how to rebuild the Model 15 TTY in the training classes offered by the Telephone Pioneers of America in the basement of Marsters' home. The classes were later moved to a Mission Street telephone utility substation in South Pasadena. Marsters frequently drove to Western Union, Pacific Bell, and AT&T plants in search of discarded TTYs. During one trip, he borrowed a large truck from a friend in the construction business, and with Weitbrecht and Slotnick he drove 450 miles to the Bay area and dropped off forty-five TTYs from the Los Angeles Western Electric plant. Each TTY measured about four feet high, two feet wide and two feet deep and weighed more than 200 pounds. Some with automatic tape sending attachments were even heavier. As a result of all that lifting, Marsters had to undergo three bilateral hernia operations.

Weitbrecht was the most changed by the TTY, however, for it had become nearly his whole universe. The technology was so woven into his life that it was difficult to distinguish the personal from the technical in his correspondence. After visiting the Taylors in St. Louis, he wrote to Breunig, "They are wonderful, most kind-hearted, and I enjoyed all the people, including the screaming children and the warm puppy . . . Paul is surely at times driven to desperation to quiet the hollering kids which now and time bothers the Phonetype."[8]

By the late 1960s, the TTY's slow spread was awakening deaf rights issues that Weitbrecht, Marsters, and Saks had addressed only indirectly. Many deaf people looked beyond their struggle to locate the old machines and blamed the Bell System as the obstacle to their access. The issue of *equity* became increasingly important as access to the used Baudot TTYs expanded. But equity meant more than may be implied in the universal service concept espoused in the Communications Act of 1934. To deaf people it meant independence, and it meant educational and employment opportunities as well.

As far as the equipment was concerned, the telephone companies were unfair in expecting deaf people to purchase their own while hearing customers were not required to do so. In Washington, D.C., a new advocacy group called the Deaf Telecommunicators of Greater Washington published a newsletter and publicly complained in November of 1968, "We should file a petition with AT&T to provide all deaf people with these machines at no cost. They (AT&T) would eventually get their money back." The group was also concerned about the issue of independence. Upset over the fact that deaf people still could not make a basic call to the telephone company itself without the assistance of a hearing person, they argued that AT&T should "handle communications between deaf and normal hearing persons."[9] A national relay service was needed, and the Bell system should take the lead in its establishment.

Without assistance from the telephone companies, the entire TTY network had to depend on community volunteer efforts and Weitbrecht's expertise at prolonging the life of discarded machines. Wearing the same baseball cap for years, Weitbrecht drove around and picked up TTYs in a Chevrolet station wagon, accumulating more than 160,000 miles on the odometer. He could diagnose a TTY problem simply by reading a description sent to him in the mail, and people around the country often wrote to him with their problems. In effect, Weitbrecht became an electromechanical "Ann Landers" of the deaf community.

But it seemed that every new TTY installation was accompanied by a new story of frustration and resistance. Telephone company personnel were not being adequately trained to deal with the new device. When installing telephones in homes, service representatives who had never seen a Phonetype modem invariably questioned whether it was legal. Deaf people did their best to describe the function of the modem and the cradle on which the telephone handset was placed. They also explained that use of the Phonetype was permitted by telephone companies and the FCC. Still, the disbelieving phone company workers often turned to their managers to discuss the matter before installing telephones.

When Weitbrecht's friend Gene McDowell requested a telephone for his house on the Missouri River about ten miles from St. Louis, he first approached the Orchard Farms Telephone Company. The staff he encountered refused to believe that a deaf person could use a telephone. He invited Weitbrecht to Missouri to help him. After Weitbrecht finally convinced the company staff that it *was* possible, a "wizened old farmerlike character" working in a general store came over to install the telephone in McDowell's home. Once the telephone was installed, McDowell then learned that it wasn't as easy to use as he had hoped. He was on a multiparty line with seven other houses, and he was expected to give his telephone number by voice each time he dialed long distance. Each time he used the telephone he had to hope that the other parties would not be confused by the sound of the TTY tones. As Weitbrecht summarized this experience to Breunig, "So goes the problems of the impecunious Independent Telephone Companies, trying to compete with the mammoth Ma Bell."[10]

During this visit to St. Louis, Weitbrecht also flew McDowell around the countryside in a rented private airplane. Inspired by a flight with Marsters, he had taken flying lessons and acquired a pilot's license two years earlier. But, oddly, as important as the Phonetype TTY coupler was to the deaf world, Weitbrecht did not include the breakthrough in a personal essay he penned later. Instead, he wrote that "probably the biggest thrill I ever had in my entire life [was] my flying an airplane on my own."[11]

Communicating with telephone operators was one of the most difficult problems for deaf people. Direct distance dialing had yet to be installed in many regions of the country. When Jim Haynes and Larry Laitinen were visiting with Weitbrecht in California, Weitbrecht decided to take them to see Mt. Wilson Observatory. In order to call first, he needed their help. He became infuriated that he could not be independent. As it turned out, such a call was not easy for his hearing friends to make either. First, they had to dial "O" for the operator. Then they had to ask for "Red Box #2." They never got through. The operator of the California Water & Telephone Company did not even know how to contact "Red Box #2." The persistent Weitbrecht showed his friends how deaf people coped with such problems in pre-Phonetype days. He *drove* them up the mountain to the observatory for a face-to-face meeting. They got in and enjoyed the day learning about solar telescopes and sunspots.

Weitbrecht also told them about how other deaf people dealt with telephone operators who broke in to request the calling number.

Cookie Williams of Wilmington, Ohio, actually had a library of tape reels, one for each person she normally called. On each tape, she identified herself, explained that she was deaf, and gave the operator her number and the number she wanted to call.

The telecommunications movement continued to grow. Telecommunications advocacy groups in various cities became increasingly successful in collecting and distributing TTYs. They also encouraged volunteers to raise funds, rebuild TTYs, and demonstrate the telephone device to others. For example, in the New York City area, Mark Bernstein, Barry Galpern, and others had volunteered to recondition TTYs. In Massachusetts, J. Thomas Rule, with the assistance of Father John Fitzpatrick, had pioneered setting up the New England Communications Service for the Deaf. He, Cary Hobson, and several other deaf men regularly carried heavy TTYs through a cow pasture to a shed in Lexington, Massachusetts, which they used as a workshop to overhaul the machines. Later, they moved to the Deaf Community Center in Framingham, and from there the organization grew slowly. Many other cities had similar groups starting up.

Deaf people with scientific training also offered their assistance. Alfred S. Marotta, a research physicist at the Department of Defense in Watertown, Massachusetts, joined Rule's group to help rebuild TTYs. In Washington, D.C., Alfred Sonnenstrahl, a deaf engineer, brought three Model 15 TTYs from New York City. One of them was given to Henry Lee "Bud" Dorsey, who formed the Deaf Telecommunicators of Greater Washington. Dorsey's group was assisted by the Telephone Pioneers of America, the Chesapeake and Potomac Telephone Company, AT&T, and Western Electric Company, which provided TTY maintenance classes on the Gallaudet College campus. Within ten months, the ambitious group installed more than eighty TTYs.

There were not that many deaf engineers or scientists in the workforce at this time, but those there were, along with other deaf people who had some electrical or mechanical skill, seemed to gravitate to the TTY movement. In St. Louis, Missouri, for example, Paul Taylor trained deaf people who did not have technical skills but wanted a tele-phone badly enough that they were willing to learn how to rebuild a teleprinter. He also teamed with Gene McDowell to publish a series of articles entitled "Understanding Your Teletype" in *The Deaf American*. The articles gave deaf people instructions on how to rebuild TTYs. Another article, entitled "Establishment of a Large Telephone Communication Network Among the Deaf in St. Louis, Missouri," provided some guidance for local groups on how to work with Western Union, the Telephone Pioneers of America, and other community or-ganizations to create ways to communicate information long distance through the TTY-telephone system.

A growing network of deaf people also saw the TTY as a way to gain access to news and weather reports. In the late 1960s, there were no captions for television; both television and radio were inaccessible to deaf people. Taylor's St. Louis group pioneered a TTY news service, a weather reporting system, and a deaf-managed relay service. The group, renamed the Greater St. Louis Telephone/Teletype Communi-cators, worked with Western Union and the St. Louis Weather Bureau at Lambert Field to transmit regular weather reports through their lo-cal TTY network. The weather reports were changed every two hours and included warnings of impending ice storms, tornadoes, sudden changes in temperature, and forecasts. All that deaf people needed to do to access the service was to call a specific telephone number. Then their TTYs would print the most recent forecast.

The St. Louis group, working with Western Union and Southwest-ern Bell, made United Press International and Associated Press news accessible in the TTY network. But the standard teletypewriter serv-ices covered only national and international news, not local events. They did not include announcements about open-captioned films, signed or interpreted lectures, or special employment opportunities that might be of interest to some deaf readers. A local TTY news serv-ice addressed many of these needs. As deaf people sent in news items, the information was typed on a punched paper tape. Then, when a special number was dialed, the news was automatically printed on the caller's TTY. The news service was, in a sense, a forerunner of the computer bulletin board, although it was a crude system of paper-tape technology.

Shortly after these news and weather services were established in St. Louis, other non-profit organizations came on the wires. These news and weather centers in the deaf community stimulated both personal and community growth and they perpetuated the TTY network. They soon evolved into important social calendars. Since only those people with TTYs could receive the news services, the printed reports

Automatic TTY punched-tape news services announced social events, local meetings, and new TTY installations, and they provided access to a variety of community services and news associated with the national TTY network. Courtesy of Telecommunications for the Deaf, Inc.

also served as a strategy for recruiting new TTY owners. Within a short time, there were about seventeen news centers in the U.S.; at least half of them operated in private homes and garages.

Missouri's first TTY answering service, a forerunner of today's telephone relay service, was not very successful. About twenty families subscribed to it for six months. But as Saks and Marsters had learned earlier in California, neither the deaf subscribers nor those who accepted the responsibility to relay the telephone calls anticipated the costs and amount of time needed for an effective service. Still, the concept of a relay service was so critical for deaf people in advancing the usefulness of TTY technology that over the next few years there were many attempts around the country to improve upon the efforts in California and Missouri. In some towns, individuals served the intermediary role in their homes. In Kansas City, Kansas, for example, a minister relayed calls for twelve families with TTYs. He assisted in everything from beauty parlor appointments to summoning ambulances. In New York City, a blind man, at no cost to deaf people, operated from his apartment. These volunteer relay services were far from adequate and frequently lasted for only a short period of time.

Even with all the technical progress made, by 1969 the TTY connected to a Phonetype modem remained a delicate instrument, requiring user skill and constant struggles with phone companies. Sometimes the telephone handset leaked electric signals, and the TTY would not print. At other times random room noise caused misprinting. Televisions, radios, screaming children, hammering, or noise from the TTY motor itself could garble phone messages in a way similar to listening to someone speak with a mouth full of popcorn. Even radio signals from nearby broadcasting stations would sometimes jumble the letters, making it necessary to install special filters in some locations.

The problem of garbled text was especially frustrating. Slightly changing the spacing between the telephone handset and the cradle sometimes reduced garble. Some people found that the problem could be alleviated by stuffing things between their telephone handset and the cradle, including sponge rubber, small wads of cloth, or even a handkerchief. The TTY sometimes printed better when the handset was completely out of the modem cradle. As the network expanded, Weitbrecht continued to provide better solutions in a series of technical bulletins he sent out. With APCOM's success depending on the acquisition of discarded machines, though, he appeared uncertain about what was to come. "Time will tell how well we succeed in our venture."[12]

Weitbrecht was the deaf community's most knowledgeable representative in dealing with the Bell Telephone system. The technical issues went far beyond the electrical and mechanical operations of the Baudot teleprinters. Weitbrecht also battled with the Bell system over another problem — modem tone loudness. Personal communication using a modem was a relatively new concept to telephone company employees, from the executive level to the personnel who installed telephones or ran switchboards. To many of these people who were unfamiliar with engineering, voice was for human communication; tones were for machines. Consequently, a deaf person's problem with tone loudness seemed to make no sense.

Usually, a hearing person needed only to speak louder if the person at the other end of the telephone conversation complained that the message was difficult to understand. But such an adjustment with a modem was not easy to make. Weitbrecht's primary concern was that reducing the loudness level to satisfy AT&T's preferences would make the signals weak, thus introducing more garble. Customers might then complain about Phonetype, not realizing that telephone company restrictions were responsible for the problem. He recruited his friends around the country to help him make measurements as he telephoned them and adjusted the transmission levels. At a later point, he considered selling a $.75 adapter for the modems already in use in the TTY network: "This would show the telephone company our interest and concern in their problem of avoiding undue noise and crosstalk from Phonetype operations."[13] The problem was not easy to solve, however; and the Bell System pressured Weitbrecht for a long time.

7

CHANGE AGENTS

During the early years of their collaboration, the APCOM partners had remarkable stamina despite pressure from many directions. Marsters and Saks were living fulfilling personal lives with a wide circle of friends and involvement in a variety of community and social affairs. James and Alice Marsters were raising three children; Andrew and Jean Saks were raising two. Weitbrecht, however, was not bearing up as well as his more well-rounded colleagues.

Vulnerable to discouragement and melancholy, he had difficulty forming close personal friendships and avoided serious relationships with women. In 1969, for example, a woman who was renting the other half of his duplex was also storing TTYs in her garage for a nominal fee. She may not have had any romantic interest in him; perhaps she was just fascinated with his abilities. But when she seemed "rather forward" with Weitbrecht one evening, he wrote to Marsters and Saks that he was worried that his "business arrangement" with her might be compromised: "She invites me in for wine, sometimes suppers, and she tries to get me to go with her and her family on various outings and sightseeing trips. At any rate, I have remained very very Victorian in my viewpoint and as far as I am concerned I intend to remain that way."[1]

He socialized very little with anyone and was awkward when he did. Rather than talking face-to-face about personal matters, Weitbrecht often preferred to write notes. He was prone to neglect his grooming, and he often surprised his partners with outbursts of pent-up frustrations, particularly about the resistance he met in his lifelong quest to be respected in scientific circles. He struggled unsuccessfully for recognition of his talents among his colleagues in astronomy and at the

Stanford Research Institute. His failure to be recognized by hearing colleagues as he thought he should be left him bitter. Despite the importance of the acoustic coupler as a technological breakthrough, he never found peace of mind with his role in the scientific community.

Yet Weitbrecht was a distant hero to many deaf people. At the Leadership Training Program in Northridge, California, Joseph Slotnick paid him a worthy tribute in a lecture to deaf leaders.

> We can all be proud of the fact that the creative genius of man is such that a deaf person, with a wonderful electronics background, and with help and encouragement, can come up with a product that is of everlastingly great use for his fellow deaf friends.[2]

But for the most part, Weitbrecht did not feel appreciated in his own small world. As he devoted more time to APCOM, this situation grew worse. He began to fall behind in his primary field of astronomy instrumentation. Yerkes and Lick Observatories were moving into minicomputers and modern semiconductors. Weitbrecht was not prepared, and in March 1969, his career at the Stanford Research Institute began to unravel. He received warnings that he might be among several scientists to be laid off. "I have been bending every effort to getting some work for SRI on some astronomy projects at Lick Observatory," he wrote to Marsters and Saks, "I am not very happy about the way SRI and Lick looks down on my 'electronic competence.' I am now 49 years old. . . . It is the old story—older people are becoming phased out of work as the younger people take over."[3]

When a potential new opportunity at the Stanford Research Institute materialized, though, Weitbrecht shunned it. John Lomax, from the Institute's Communication Techniques Laboratory, suggested to Weitbrecht that he consider joining the laser research project because they needed someone with optical experience. Weitbrecht resisted. On April 25 he wrote Lomax, "I wonder as to how to comment on this in a positive direction. I have problems of an interpersonal nature."[4] He was concerned about his age and his deafness. He told Lomax that he was more interested in computer operations and asked him for suggestions about "where I may best fit in." Lomax warned Weitbrecht about his reluctance to consider the Electromagnetic Techniques Laboratory project: "If you choose to not apply yourself wholeheartedly to this work . . . then I do not see how I can assist you any further here at SRI."[5]

This was not a good time to be uncooperative at SRI. One of the nation's largest nonprofit research organizations, the Stanford Research Institute was experiencing growing discontent among Stanford University students, who were calling for the university to sever its ties with SRI and demand that SRI reimburse the university millions of dollars. The object of greatest campus controversy was the counterinsurgency studies for Southeast Asia. One of the younger hearing physicists conducting research at SRI on magnetic levitation, or "Maglev," had poked fun at the student unrest with a printed card, "Maglev, Not War." But the troubles were serious, and revenues had been dropping as federal support for basic research declined.

Along with this turmoil, Weitbrecht had not gotten far in his efforts to secure his rights to the Phonetype patent. The Stanford Research Institute still argued that it had some rights to Weitbrecht's creation, and Marsters was not optimistic about APCOM's future in a struggle over patent rights. He realized that the Weitbrecht modem patent would be costly to defend. He told Weitbrecht, "we have been fortunate that the ugly head of competition has not [been] raised . . . and given us a rough time . . . but I expect it will in the future."[6]

On top of his employment woes, Weitbrecht's mother had a stroke. His brother George had re-enlisted in the Air Force, leaving all responsibility on his deaf brother's shoulders. Weitbrecht therefore needed to find a place for his mother nearby so that he could take care of her. "I have been supporting Mother for many years . . . sending her monthly support money," Weitbrecht told Marsters.[7]

At about this time, he received some recognition that lifted his spirits—a citation for Meritorious Service from the President's Committee on Employment of the Handicapped. It was also at this time that a solar eclipse was approaching, and Weitbrecht was able to observe it in Mexico. In his own darkened world, Weitbrecht needed the escape such a trip would provide. Perhaps the brief expedition soothed him in this time of anger and resentment. In his youth he had taken walks at night with his family, using a flashlight to study maps of the constellations. In 1930, he had written a high school class composition about a solar eclipse. A few years later, at the age of 18, he won the Bausch and Lomb Honorary Science Award for his construction of a reflecting telescope using a Pyrex disk for a mirror and an old automobile axle for the mounting. He had long wanted a better telescope, but he just could not afford it.

He became increasingly concerned about his financial status. On May 18, 1969, he wrote to Marsters and Saks in desperation:

> I need to beef up my meager financial resources in hopes of achieving relative retirement security. And I have a sense of responsibility to our Applied Communications Corporation in connection with its efforts to provide useful communications devices for the deaf and also to penetrate the commercial market. Already I am serving two or more masters at the same time, frittering away my daily energy in pursuing unrelated activities. Is that good? . . . Both of you are a "light in my life," and I am thus inspired and encouraged to do my best not only for Apcom but also for the deaf world.[8]

In late May, the situation grew worse. Weitbrecht was laid off by the Stanford Research Institute. He was crushed, "Truly, this is one of the most TRAUMATIC events I have ever had in my life . . . I just feel that my reputation, let alone interest, as I may have had in science, is shattered; and it will be rather difficult for me to recover any such."[9]

James and Alice Marsters, concerned about Weitbrecht at this critical point, took him and his mother in for several weekends in a row. Weitbrecht also momentarily distracted himself by working with Paul Taylor on visualizing the "ESSA-Radio-Phonetype-Severe-Weather-Warning-System" for St. Louis. Anxious to find a new source of income, he immediately asked for a salaried position at APCOM, but the company did not have the capital to hire him on a long-term basis. In particular, Marsters and Saks saw no way the not-for-profit company could offer Weitbrecht retirement benefits.

Weitbrecht looked at the matter somewhat differently, though. He knew that APCOM would collapse without his help. In fact, he did not mind pointing out with extreme candor the lack of technical design skills in his partners, and he spent an inordinate amount of time detailing his experiments to them in a constant flow of correspondence. Finding another research position in industry did not appeal to Weitbrecht. He wrote to Marsters and Saks:

> I am simply disgusted with these 'TANKS of Engineers and Scientists,' and I am very, very reluctant to consider going to work for big outfits. I just do not obtain recognition in such environments, and I feel my communication problem works to my disadvantage as far as worthwhile projects, promotions, and job-satisfaction are concerned. . . .
>
> Both of you . . . are blessed with family fortunes and hence do not have to worry one whit as to where the money will come from. I have practically nothing, and what monies I get must be earned through my own sweat. I hope you understand my problem. Even just now, my Mother has asked me to cash in her 50-odd shares—all she has in the way of resources.[10]

Marsters fruitlessly encouraged Weitbrecht to consider "a grant to do some work which is not lucrative to us, but which will benefit APCOM indirectly, as well as the handicapped people."[11] But after his recent employment experiences, Weitbrecht was not interested in grants and was disdainful of scientific research project teams. "I guess I am thinking a lot about private enterprise, and need for encouraging more of such, rather than depending on government handouts."[12]

Through the summer of 1969, Weitbrecht put aside his personal and financial problems and continued his quest for a more portable TTY. He met with his ham radio friend Don Goodwine from Guadalajara, who gave him some ideas on miniaturizing the Phonetype terminal unit. Goodwine's lightweight MITE (Miniature Integrated Teleprinter Equipment) Corporation Page Printer weighing less than ten pounds, interested Weitbrecht. Upon returning to APCOM, he wrote to MITE, "We want to work with you, if possible, to develop a low-cost

Mite Corporation's TTY was lighter and smaller than most other TTYs in the late 1960s. The Applied Communications Corporation used this machine for demonstrations around the country. Reprinted with permission of the *RTTY Journal.*

Baudot version of the Model 123 system with specific application towards use by deaf people in their homes, offices . . . and other places."[13]

Weitbrecht was not the only person interested in developing a portable TTY. In the fall of 1969, Ross Stuckless, director of the research department of the National Technical Institute for the Deaf at the Rochester Institute of Technology (RIT), encouraged Jim Walker to consider developing a portable TTY and provided modest support to cover some costs for the research. In a few months, Walker, chairman of the RIT Department of Electrical Engineering, and another member of his department, George Thompson, had two prototypes ready for testing. Each weighed about ten pounds. After a field test between California and Rochester proved successful, Walker and Stuckless thought it was time to try it out with deaf people. They contacted the National Association of the Deaf Executive Secretary Frederick C. Schreiber and Boyce R. Williams at the Vocational Rehabilitation Administration.

Walker and Stuckless flew to Washington with the two prototypes and a repair kit should the need arise. Checking into a hotel, Walker asked Stuckless to test the devices from their respective hotel rooms. The prototypes did not work. Walker disassembled the TTY, found nothing wrong, reassembled it, and asked Stuckless to try again. Back in his room, Stuckless called again, but once more it failed to transmit. By now, it was midnight. At his wit's end, Walker asked Stuckless if he had placed the telephone handset on the acoustic coupler backwards. That was the problem.

The demonstration call between Schreiber and Williams went well, and Stuckless and Walker returned to Rochester to continue developing the prototype. Within a few months, however, they heard that a portable TTY was being designed elsewhere. They halted their research prematurely. It would be three more years before such a device would come onto the market.

While research on the portable TTY was going on in 1969, the TTY network found a strong advocate on the East Coast—I. Lee Brody, although he would eventually cause problems for the APCOM partners. Ever since Weitbrecht had installed the TTY station at William Bernstein's house in New York City, he had wondered why the local network did not expand more quickly, the way it had in St. Louis. In 1968, about twenty machines were reconditioned and installed in deaf people's homes in the New York area. Recently, Brody had begun to expand this local network.

The deaf son of a tavern owner who had emigrated from Austria and set up business in Newark, Brody grew interested in the TTY movement after a harrowing personal experience. While hunting in upstate New York, he slipped on some wet rocks and was paralyzed from the waist down for about six hours. He was unable to get help by firing distress flares and his shotgun. During that experience, he did a lot of thinking about deaf people in emergency situations. Slowly, he began to recover some sensation in his legs and he was able to crawl and stumble for several miles until he reached safety. Later, while in the hospital for a spinal fusion, he learned about another deaf man who had had a heart attack, and his wife hadn't been able to find help in time. These incidents made him think a great deal about telephone access.

After Brody left the hospital, a friend hooked up a Phonetype coupler to an old Western Union TTY. Brody was fascinated with it and in the spring of 1969, he set out to form New York-New Jersey Phone-TTY in Fair Lawn, New Jersey. He learned that most TTYs in that region had been coming from ITT World Communications. At that time, the company had a list of sixty-five deaf people waiting for teleprinters. He sent a message to TDI's Breunig stating, **I AM BEGINNING TO THINK THAT WAITING FOR A TTY AT THE END OF A RAINBOW IS GOING TO BE A LONG WAIT.**[14]

In response, Brody placed advertisements for TTYs in such magazines as the *RTTY Journal* and *RTTY Bulletin*, and he answered ads placed by ham radio operators. In newspapers and TV news programs, he publicized the purpose of the network with the hope of attracting the interest of those who were able to offer TTYs for reconditioning to his local group. Usually, only one or two machines could be obtained at a time. Brody also spent many hours with Western Union and Bell Telephone service representatives in order to learn firsthand about rebuilding TTYs. During the first year of his work, the Telephone Pioneers of America provided technical instructors to train deaf people for fifteen weeks at the New Jersey Bell Training School in New Jersey.

Like Weitbrecht, Brody tried hard to convince AT&T to manufacture a special TTY for the network. He wrote to Teletype Corporation Product Development, "Is it possible for your company to manufacture a new page printer without any internal wiring . . . to sell for around $100 or near that figure?"[15] The answer to Brody's question was clearly no. Whether or not Bell Laboratories was capable of developing a portable TTY in 1969 is another question, but in any case deaf volunteers would have to continue rebuilding the older machines. Brody would continue looking for better solutions, leading to a nearly disastrous split within the deaf community.

Meanwhile, Weitbrecht, the volunteers' technical leader, did not find a new job after leaving SRI. By midsummer of 1969, he decided that he wanted to work full-time at APCOM, even though the position would not include retirement benefits. "As for my salary," he wrote to Marsters in late July, "whatever you and Andy have decided upon is very fine by me. APCOM is certainly very kind to me, and I want to show worthwhile results."[16]

At APCOM, Weitbrecht worked with volunteers and TDI agents, the individuals given responsibility for distributing TTYs to deaf people

I. Lee Brody, founder of the New York-New Jersey Phone-TTY organization, reconditioned TTYs in his home in Fair Lawn, New Jersey. Courtesy of NY-NJ Phone-TTY, Inc.

and instructing recipients in their operation. He prepared and mailed technical bulletins to the agents with such titles as "Teleprinter Machine Connections for Use with Phone-TTY System," "Ordering Telephone Service for Phonetype Operations," and "Adjustments in the Range Finder-Selector Magnet Area." He exhorted more technical help from deaf men and women, calling on them to join the movement. Emphasizing a recurrent theme in his own thinking, he wrote to his partners "Deaf people need to be challenged to exert greater interest and effort in improving their own lot."[17] He and Marsters then followed through by assisting TDI in demonstrating TTYs in the hope that others would join them.

Saks and Marsters also experimented with an early form of voice carry over (VCO). Weitbrecht later described the creative technique as an "accessory to permit a deaf person to talk directly to a hearing person in voice, and the hearing person would type back to him. . . . on a single telephone line."[18] These first efforts, although unsuccessful, nevertheless firmly established the necessity for both relay services and voice carry over and served as catalysts for others to experiment with them in the years to come.

Attitude problems continued to limit the TTY's spread and usefulness, the APCOM partners believed, so in August 1969, Marsters and Weitbrecht embarked on a long journey. They traveled 8,000 miles around the country with a lightweight MITE Corporation teleprinter and a Phonetype in a demonstration van. There were about 600 TTYs in the national network by then. The formation of local telecommunications groups had been helpful in spreading the word about the TTY's utility, but it was not easy to change attitudes. Weitbrecht and Saks thought many deaf people, resigned to accepting less from society than their hearing peers, had become accustomed to life without the telephone, and they just did not realize its advantages. The demonstrations attempted to address that.

TDI was experiencing similar difficulties with attitudes and habits deaf people had developed from lifetimes of protecting themselves. TDI believed that a national directory of TTY users would be valuable, but initially the organization encountered difficulty in convincing TTY owners to list their phone numbers. Various reasons for their reluctance were given, but a significant one was fear of exploitation. Some worried that being identified as deaf in the national directory, now known as the "Blue Book," might increase the chance that burglars would target their home, assuming that deaf people would not be aroused by noise.

Weitbrecht and his partners thought that another approach to changing attitudes was through recognized deaf community leaders, and they began an effort to place loaned TTYs in their homes. One TTY, for example, was loaned to Leo Jacobs, a deaf activist and mathematics teacher. After a trial period of six weeks, Jacobs asked Weitbrecht if it were possible to obtain twenty more Phonetypes for his friends. Weitbrecht sent a memo to his partners: "Thus you can see the need for making loans to selected people and trying to get them hooked."[19] Later, after Jacobs telephoned his daughter in California, he wrote of the "undiluted joy that I was able to hold an unsupervised conversation with my family."[20] Although at first this appears to be a simple emotional response, the notion of "unsupervised conversation" was a poignant interpretation of the implications of this breakthrough.

The strategy of recruiting deaf leaders worked well. Fred Schreiber, executive director of the NAD, had been given a Weitbrecht modem. He convinced Malcolm Norwood to set up a TTY in the U.S. Office of Education. Norwood and another deaf man, Ed Carney, then supplied TTYs to the three major distribution centers for captioned educational films. In time, these efforts led to more than sixty other centers acquiring TTYs. The telephone access by deaf employees in these centers made the captioned educational films program more efficient. It set an example for other organizations and agencies whose business depended on long-distance communication.

Deaf leaders often made tremendous personal sacrifices to demonstrate the effectiveness of the TTY. After Alfred Sonnenstrahl and Henry Dorsey delivered the first three TTYs to the Washington, D.C., area, Dorsey lugged one of the heavy Model 15 TTYs to one deaf family's home after another. Within a month of his effort, the TTY network increased to ten. These deaf leaders were the gatekeepers who channeled information and used creative strategies to help weave the technology into deaf people's day-to-day lives. But even the leaders were often slow to accept the TTY. Schreiber at first saw the potential only for organizations. He did not think the machines would be popular in the home.

Even Gallaudet College in Washington, D.C., with its long tradition of providing quality education to deaf people and with an enrollment of hundreds of deaf students and many deaf faculty and staff, had only one TTY in 1969. It was in the chaplain's office. A recent survey about telephone access had shown that the deaf faculty and staff were more interested in the expensive Electrowriters. Sonnenstrahl, a young man then in his mid-20s living in Maryland, brashly assumed he could do something about it. He telephoned the chaplain's office by TTY and asked for Gallaudet President Edward C. Merrill. To answer the phone, Merrill had to walk across the campus in a driving rainstorm. Once he arrived at the chaplain's office, he greeted Sonnenstrahl, who asked him, **HOW DO YOU LIKE THIS INCONVENIENCE?**[21] Merrill proceeded to order eight new Model 32 TTYs to be installed on campus. Unfortunately, only the faculty and staff had access to them. The students had to wait another year before more TTYs were ordered.

Weitbrecht followed the growing TTY network with both pleasure and concern as he recognized new problems. Sonnenstrahl, for instance, was paying several hundred dollars per month for long-distance telephone calls, reminding Weitbrecht that something had to be done about rates for TTY users. But he was too bogged down with correspondence about TTY maintenance problems to address this issue. Not only were the bulky TTYs an interior decorator's nightmare, there were also many different models. Teletype Model 15 and Model 19 machines were very common. Since the 1930s, about 300,000 of them had been manufactured. Both became obsolete when faster TTYs were introduced. In one letter, Weitbrecht explained to a deaf man that the complications he was experiencing with his huge Model 19 TTY required APCOM to draft a special bulletin on the wiring in that model. In another TTY conversation with Ed Carney in Washington, D.C., Weitbrecht tested various messages and determined that his Model 32 TTY required disabling the "answerback device." Carney had reported that when a symbol such as the dollar sign ($) was typed, the TTY invariably acted **LIKE A LOS VEGAS SLOT MACHINE THAT HIT THE JACKPOT.**[22]

Deaf people who wanted to use the telephone on a regular basis either had to learn technical jargon and maintenance strategies themselves or suffer long periods without service while waiting for a TDI agent to come and service the machine. The discarded TTYs were often in deplorable condition when they were received from AT&T. Often, a deaf person would not know whom to contact when a TTY malfunctioned. With many possible causes of garbled messages, including dirty keyboard contacts and telephone handset shape, service representatives from the local organizations were continually being called to troubleshoot. In one call to Weitbrecht, Taylor emphasized how it had taken him nine hours to adjust the tapefeed sprocket and relay bank on one TTY.

In some cities, hearing people took the lead in establishing the local telecommunications groups. The Indiana Deaf Communicators started with two hearing men who, because they were married to deaf women, were well aware of the telephone access problem and helped recondition and distribute TTYs to households. Soon after, deaf people joined them on a maintenance team they established. Hearing men also set up groups in Pennsylvania and Florida. Although TDI recognized the important role of hearing "coordinators," the national organization reserved the title "agent" for deaf persons in the interest of fostering self-empowerment.

Support of the TTY distribution program through TDI appeared to be the only solution to telephone access for deaf people. Volunteerism across the country reached an unimaginable level. TDI agents spent evenings and weekends coordinating the search for TTYs, their storage, reconditioning, and installation in homes. When surplus machines accumulated, storage became a problem. One weekend, AP-COM was to pick up thirty TTYs from Pacific Telephone and Telegraph. Seventy-five of the bulky machines were already in storage at the California School for the Deaf at Berkeley, and the school was concerned about space. Weitbrecht had twenty-two in his own garage and was paying his neighbor ten dollars per month to hold twenty more for him. Other towns had none. When the New York-New Jersey Phone-TTY group had trouble finding enough TTYs to meet the demands, Rev. J. Edwin Hewlett of St. Louis arranged with Pacific Intermountain Express to ship eighteen TTYs from Seattle, Washington, to New Jersey at no charge.

The following year, Rev. Hewlett shipped forty-eight TTYs from Berkeley, California, to Washington, D.C., with contributions from his congregation. Then deaf people in his own hometown ran out of machines. "Better tell your friends to get a teletype right away," the St. Louis group warned its members, "before they sadly realize that there are no more teletypes to be given away in the very near future."[23] Marsters personally delivered discarded TTYs to his friends. Nearly every time he flew to the East Coast on commercial airlines, he took a teletypewriter in two cardboard boxes to give to someone. He suggested this strategy to others, but few were willing to carry the heavy machines.

Deaf women also became involved in the movement, especially as officers of local groups and as participants in fundraising efforts. Sally Taylor was one of the first women to provide such support. She assisted the Telephone/Teletype Communicators of St. Louis through record-keeping and correspondence, served as editor of the first TTY news service for deaf people (the *St. Louis Messenger*), and helped plan social events to raise funds for the TTY movement in Missouri. When her husband Paul installed machines in homes in St. Louis, he often called Sally at home. First came the test message "RYRYRY." Then came **THE QUICK BROWN FOX JUMPED OVER THE LAZY DOG'S**

During the late 1960s, Henry Lee Dorsey of Silver Spring, Maryland, rebuilt discarded TTYs, which allowed deaf people in the Washington, D.C., area to use the telephone. Courtesy of Telecommunications for the Deaf, Inc.

Like many other deaf women, Sally Taylor was deeply involved in the movement to bring telephone access to deaf people. The Taylors often had more than thirty teletypewriters, each weighing over 200 pounds, in their basement, waiting to be reconditioned. Courtesy of Sally A. Taylor.

BACK   1234567890 TIMES (which comprised all five segments of the Baudot code and was helpful in adjusting the range finder of the old TTYs) to test all the alphabet and numeric characters on the keyboard.[24] At one time the Taylors had three dozen TTYs the size of drop mailboxes in their basement waiting to be reconditioned for their friends.

Deaf women also served as authorized TDI agents responsible for reconditioning and distributing TTYs. Of about forty authorized agents recruited by TDI in the first four years of the network, two were women. A half dozen more became agents over the next few years. Among the many deaf women volunteering with the TTY movement was Ethel Johnson, who helped TDI with the mailing of directories, newsletters, and correspondence. She typed and mailed all of the affidavits required to be signed as part of the agreement with AT&T. Yet Johnson herself did not have access to the telephone at home. Her neighbor in the apartment below had complained about the TTY noise, so she was asked by her landlord to discontinue its use.

In APCOM, Weitbrecht was becoming increasingly aware of developing technologies in the area of electronic visual displays. He found one article in *Control Engineering* magazine about a device being developed to replace the Model 33 ASCII Teletype. The new equipment was aimed at the timesharing computer market, and he knew APCOM would be left behind. The writing was on the proverbial wall, showing the way to various forms of electronic visual readout devices.

September passed quickly. During the first weekend of October, Weitbrecht prepared five Model 15 TTYs to distribute in Oakland. He put on a demonstration at the Lutheran Church for the Deaf there, hoping that the fifty pastors in that group would want Phonetype installations. APCOM also loaned two TTYs to the California School for the Deaf at Riverside. The partners believed that regular use of the TTY by deaf children might improve the children's reading and writing skills. The school in Berkeley had already borrowed two TTYs, and the machines had become so popular that it was necessary to ration telephone time to the students.[25]

Brody's competition with APCOM began in November of 1969, when he started exploring the possibility of a less-expensive modem. He wanted to subcontract with APCOM to manufacture the Phonetype coupler in the East, but APCOM had already subcontracted with several companies and had not been pleased. APCOM had discovered that these companies tried to increase their profits by cutting corners in circuitry components, leading to faulty modem performance. Frustrated, Brody approached a company called ESSCO Communications that had been in business for years as a manufacturer of equipment for radio hams. There he met a young engineer named Jim Steel, who was willing to build an experimental new modem. Brody then took the prototype to the New Jersey Bell Telephone Company Training School to have it evaluated.

Weitbrecht soon began receiving reports that other modems were being designed and sold for less than the Phonetype. He worried more about the problems of incompatibility than financial loss from direct competition, wondering whether a lack of quality control might frustrate deaf people and discourage them from acquiring telephones.

Meanwhile, the competition for scrap TTYs and parts had become a nuisance for everyone. TDI agents were cannibalizing machines to get others to work. Radio amateurs were both grabbing up TTYs for their use and helping deaf people with telephone access. *RTTY* published a description of TDI's operation and the fundamentals of the TTY terminal unit, encouraging hams across the country to donate TTYs to the tax-exempt organization or to volunteer to train deaf persons in reconditioning and servicing them. Gale H. Smith, Assistant Engineering Manager of the New York Telephone Company, also published an article in the amateur radio periodical *QST*, describing various telephone devices being used by deaf people and encouraging radio amateurs to become involved by doing community service work or by contacting TDI to offer assistance in reconditioning the TTYs.

In March 1970, the Governor's Committee under Ronald Reagan selected Weitbrecht as one of the ten nominees for the Outstanding Handicapped American of the Year for 1969. Weitbrecht also finally received the patent for his modem. He was euphoric. To him the patent was an issue of self-esteem in the scientific community. "I felt happy to be able to demonstrate my electronics competency to the hearing world, by showing a patent as proof. After all, so many hearing people tend to look down on deaf people as being helpless and second grade morons."[26]

One thing was certain to Weitbrecht—attitudes were harder to change than anything else. Just as he and his partners were feeling good about the patent, along came a letter from another overly protective hearing person whose deaf father had independently purchased a Phonetype and had it installed in his apartment. The young man explained that deaf people in Illinois communicated with other people out of town by telegrams or letters. For local conversations by phone, they had hearing people like himself to make calls for them. He considered it "very foolish" of his father to throw away his hard earned money "for nothing" and asked Weitbrecht for a reimbursement.

Much more serious was the attitude of telephone company officials, who still, in 1970, seemed unable to understand deaf people's desire for a convenient service. "One Policy, One System, Universal

Service" was a slogan AT&T had repeatedly published for more than a half century. In May, AT&T summarized available choices in a "Chart of Telephone Equipment for the Totally Deaf." The chart included the Sensicall, Code-Com, TWX Service, Telewriter, Watchcase Receiver, Picturephone, Facsimile, and the teletypewriters distributed through TDI. AT&T listed the pros and cons for each choice. For the Sensicall and Code-Com, parties on both ends had to know a code; communication was very slow. To use TWX, the Telewriter, and Facsimile, both parties had to subscribe to AT&T's Data-Phone Service; communication was very limited. The AT&T Watchcase Receiver was only for deaf people proficient in speech. Picturephone, as stated on the chart, "will not be widely available for many years."[27] That left only the TTYs supplied by TDI as a feasible choice.

It had been five years since Weitbrecht's first meeting with AT&T representatives in New York City, and he was frustrated and angry that the company was still promoting the use of Morse code. Code-Com converted electric signals into mechanical vibrations of the finger plate and into flashes of a small lightbulb. It could only be used with a code, most often the *no, yes-yes,* and *please repeat* system, or, less frequently, Morse code. In the Institute of Electrical and Electronics Engineers periodical *IEEE Spectrum,* an article highlighted the Code-Com as a "new telephone that will allow the deaf to 'see' messages in coded flashes of light."[28] AT&T advertised Code-Com in many popular magazines. Each time Weitbrecht came upon such an advertisement, he became curious about the Bell System's intentions. He wrote to Saks and Marsters, "We think that a lot of deaf people will be rather disappointed . . . should they give it a try."[29]

TDI's Breunig also wrote to AT&T to express concern about their seeming inability to understand what deaf people wanted. He noted that there was a lack of acceptance of the Code-Com set in Indianapolis and that the printed word was the most effective medium. Clifton R. Williamson, AT&T assistant vice president, responded to Breunig that AT&T saw the advantage in the inexpensive device enabling deaf people to receive calls from anywhere in the world: "Our interest is not in promoting CODE-COM sets for the deaf," Williamson continued inexplicably, "but in providing one more way to make it possible for them to communicate by telephone."[30]

Even some of the hearing Bell Telephone system technicians shared the views of Weitbrecht and Breunig. Carl Argila had developed a special interest in the needs of deaf people and was instrumental in bringing several Phonetypes to the Philippines for the first trans-Pacific TTY call. In a memo to Bell Telephone Laboratories scientists and

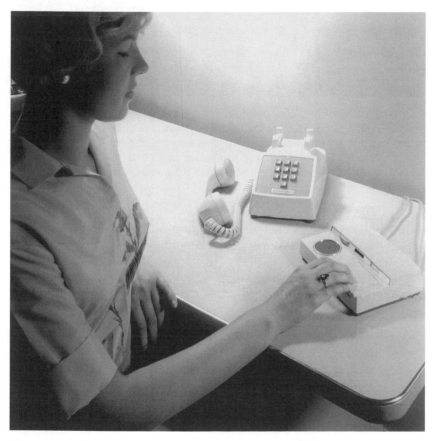

The Code-Com set was another in a long line of devices developed by Bell Telephone Laboratories for use by deaf people. It had three components, a flashing light in the center, with a circular vibrating pad on the left and a sending key, used like a telegraph key, on the right. Property of AT&T Archives. Reprinted with permission of AT&T.

On July 31, 1970, the first intercontinental TTY call was made between the National Association of the Deaf (NAD) and the Philippine Association of the Deaf in Manila. At this historic call in Minnesota were *(standing left to right)* Gordon L. Allen, Albert Pimentel, Frederick C. Schreiber, Edward Carney, and Boyce R. Williams, and *(seated left to right)* Jess M. Smith and NAD President Robert O. Lankenau. Courtesy of National Association of the Deaf.

other professionals, Argila summarized non-Bell System efforts and championed the APCOM Phonetype coupler. He recommended that the Bell System become more involved with the TTY.

> The author feels that, in the past, the Bell System probably could have responded more fully to the needs of the deaf telephone user. Services such as teletypewriter and facsimile have been offered to the deaf, but not adapted for the deaf. Such services are offered at the commercial rates, making them nearly unobtainable for even the well-to-do deaf person. Furthermore, the only unit developed explicitly for the deaf, the CODE-COM® unit . . . appears to be of practical use only to a minority of the deaf population.[31]

Argila told the Bell system that, with regard to the older TTYs, deaf people were generally reluctant to "bring into their homes such large quantities of equipment which they themselves must maintain and repair." He described the advantages of Weitbrecht's Phonetype modem, a "sophisticated analogue of the Bell System DATA-PHONE® unit," and how he had constructed a duplicate at an estimated cost of thirty dollars for parts. Based on a survey of eleven deaf people with an estimated average monthly phone bill of thirty-five dollars, he explained that the Bell System was collecting lucrative revenue "it otherwise would not have seen." Argila reported, "If one out of every ten persons with no usable hearing acquired a teletypewriter, the resulting market of 200,000 such installations would net the Bell System approximately $7 million in revenue annually, not including revenue collected for any special equipment." He recommended that Bell Telephone Laboratories manufacture and lease a compatible TTY unit.[32]

Again, a champion for appropriate telephone technology for deaf people could be found within the Bell system, but again a voice went unheard.

8

THE MODEM WAR

In October 1970, ESSCO Communications began marketing the ATC-2, a second modem for the TTY network. The following month, *The Silent News*, a national newspaper for the American deaf community, brazenly announced, "N.J. Firm Patents TTY Terminal Unit; Costs $100 less than Rival." The Modem War had begun.

Marsters sensed trouble when advertisements for the ATC-2 coupler included claims of superiority over Phonetype. In the months to follow, he and Lee Brody, who was marketing the ATC-2 coupler, had numerous angry exchanges about the advertisements. Both men had independent incomes and were motivated primarily by humanitarian interests. Yet good intentions in no way lessened the animosity created through the advertising by APCOM's competitor. Marsters fully understood Brody's concern about the cost of Phonetype, but he believed that the ATC-2 coupler had not been tested properly. Furthermore, even with the more expensive Phonetype, there were not enough TTYs to go around. What would happen if there were a flood of low-cost modems with no teleprinters to use with them?

This new development tugged Weitbrecht in opposite directions. On the one hand, he was committed to universal telephone access for deaf people and therefore welcomed lower cost equipment. On the other, he needed income, and the ATC-2 coupler's competition potentially threatened his job. Confused by these conflicting needs, he frequently sought advice from APCOM's lawyer, Reed C. Lawlor. He discussed instituting a patent infringement suit against ESSCO Communications, but he was not encouraged to do so. At one point, the attorney wrote that documents from the FCC or others might "appeal to your humanitarian interest [and] lead you to give something for

116

nothing."[1] Marsters and Saks warned Weitbrecht that the costs of instituting a patent infringement suit against an established company would put them out of business completely.

Weitbrecht also worried that the proliferation of different modems would create quality control problems in the network, which was already a significant issue because of the variety of types of old TTYs in use. Lawlor advised the deaf physicist to "try to avoid overreacting" to these concerns.[2] Exhausted and fragile, Weitbrecht would not calm down about the competition, however, and Marsters became worried that Weitbrecht's support for the TTY effort might cease. Marsters told Brody, **WE ARE CONCERNED FOR HIS HEALTH, TOO, AND WE ARE TRYING TO HELP HIM ALONG . . . SO HE WILL KEEP GOING INSTEAD OF DRIVING HIMSELF INTO THE GRAVEYARD.**[3]

By the fall of 1970, Weitbrecht's fatigue was obvious. He blamed his problems on APCOM's small staff and the competition, but he also worried about his own skills. In a letter to TDI's Breunig, he explained that

> we don't have a full-blown electronic production shop—only a tiny laboratory in an office room measuring some 12 ft. by 25 ft. . . . We three . . . have been working extremely hard in our business, trying to promote and improve a useful communication system for the deaf. And it disheartens us to be aware that the competition is pre-empting our years of effort, expense, and devotion spent [pioneering] the Telephone-Teletypewriter System for the Deaf.[4]

He was also concerned about expanding his competence in electronic design and told Breunig he was having a difficult time keeping up with the field, considering the fact he had to wear many hats in the APCOM business.

Much of Weitbrecht's time was filled with letter writing to APCOM's lawyer. Stunned and having trouble sleeping, Weitbrecht sat at the typewriter for hours on end, composing memos and personal letters. He had not taken an extended vacation for a long time. On December 6, depressed, he called Marsters. Ever generous, Marsters offered to plan a trip for him to England. Weitbrecht refused and responded bitterly:

> IT SEEMS THAT I AM RESPONSIBLE FOR THE SUC-
> CESS OR FAILURE OF APCOM. I AM CARRYING QUITE
> A LOAD. I AM THE 'SO-CALLED INDISPENSABLE
> MAN' AND INDISPENSABLE MEN HAVE A WAY OF BE-
> COMING EXPENDED.[5]

The end of 1970 found Weitbrecht alone with his dog, Bongo, taking long walks and thinking about his battles. He would have to carry the burden of outlining the Phonetype's technical details for APCOM's lawyer, who was following up on his request to examine the possibility of infringement on the patent. In February 1971, he received an honors citation from the A. G. Bell Association for his work on the TTY modem design, but this award apparently did not alter his mood. Brody sent him a Western Union telegram: "Congratulations and best wishes Bob. Today you are honored as Mr. PTTY—the man who has brought light and hope to the deaf community of the world."[6] Weitbrecht was so deeply offended by his competitor's behavior, however, that he waited until April to send a thank you. Even then, he included a comment about how piqued he was by Brody's tactics in promoting the ATC-2 coupler.

In the midst of these problems, the APCOM partners succeeded in getting the federal government's attention. Several years earlier they had written to the Internal Revenue Service (IRS) and requested a deduction of the cost of a Phonetype modem as a medical expense. At first, the IRS had refused, saying that eligibility of a medical expense deduction depended on it being ordered or directed by a physician. "This letter is preposterous," Saks wrote to Marsters.[7] Each of the APCOM partners then mailed his own letter to the Commissioner of the IRS, requesting a deduction in the cost of a Phonetype modem.

The IRS took the matter more seriously this time. It investigated the situation and ruled that the acoustic telephone coupler was deductible as a medical expense. This tax deduction was the first formal

(Left) APCOM's Phonetype acoustic coupler was first marketed in 1967. The prototype was developed in 1964. *(Right)* ESSCO's ATC-2 acoustic telephone coupler was marketed in 1970. From the collection of I. Lee Brody, courtesy of NY-NJ Phone-TTY, Inc. Photographs by George Potanovic, Jr./Sun Studios.

recognition of the communication needs of deaf persons by the federal government. TDI published the news in its quarterly newsletter: "All deaf persons who are entitled to this deduction owe a debt of gratitude to the officers of the Applied Communications Corporation who made application for this ruling . . . Take a bow, Jim Marsters, Andy Saks and Bob Weitbrecht!"[8]

The partners' success with the IRS ruling suggested that activism might be more effective than volunteerism in the fight for equitable telecommunications services and products. That feeling was starting to spread to the local telecommunications groups. One example can be found in the first annual report of the New England Communications Service for the Deaf, which stated, "In this day and age of technological advancement, the deaf are finally arriving at a place whereby they can begin to make themselves heard and their needs known to others besides themselves."[9]

APCOM, TDI, and the National Association of the Deaf increasingly saw deaf people's needs as public concerns, appropriate for government action. The federal government had the ability to provide compensatory measures for people seeking visual access to the telephone, but it had not exercised this power. This passivity led to troubling questions. Why were the needs of deaf persons being neglected and numerous opportunities allowed to pass by? Why weren't technologies transformed into appropriate assistive devices? Couldn't scientific research and its technological applications be under the democratic control of society? Deaf people seemed to have gone through a long period of acquiescence to technology controlled by and for *hearing* society.

Bolstered by success with the IRS, Marsters approached AT&T again, this time about combining their research and development efforts to help deaf people gain equitable services. On April 20, 1971, he wrote to Ronald S. Callvert, AT&T Director of Public Relations.

> In the past, the Bell System may have been criticized for developing some expensive engineer's dreams rather than developing practical devices within the means of handicapped people. A "Bell Laboratory for the Handicapped" could help overcome this criticism and also provide necessary information on telephone devices for the handicapped to the employees of the Bell System, Associated Telephone Companies, and to the public.[10]

Months passed without a response from AT&T; meanwhile Marsters and Weitbrecht continued to argue with Brody about evaluating each other's modems and how best to ensure compatibility.

By early September of 1971, AT&T seemed to be moving in a more positive direction. Weitbrecht received a modified Model 32 TBK teletypewriter from Teletype Corporation. The AT&T subsidiary had finally produced a new machine specifically for use by deaf customers, but the engineers did not follow Weitbrecht's instructions. The machine was wired incorrectly. Weitbrecht fixed it himself and wrote a letter to Teletype Corporation to straighten things out. Even with this problem, he was happy to have this new TTY with an automatic carriage return and linefeed, although it cost $520. Meanwhile, ESSCO announced that Scan-a-Type, a new solid state digital teleprinter, would soon be on the market.

Marsters continued his efforts to convince AT&T to become more involved in the TTY movement. In November he met with Ronald

Callvert in New York City and followed up with a letter that again suggested a merger of sorts.

> It is proposed that the Bell System acquire the Applied Communications Corp. as a source of increased revenue, low-cost public relation possibilities; information and demonstration center for telephone installers and the public of special equipment; and to insure on-going low-cost research and development at the grass-roots level of communication devices for the handicapped.[11]

Based on his previous experience, Marsters knew AT&T would not accept the logic of his proposal. The company still appeared unable to see the revenue benefits that deaf telephone access promised.

On November 13 and 14, 1971, the First National Conference of Agents of Teletypewriters for the Deaf, Incorporated, provided a different form of optimism. The Office of Public Service Programs at Gallaudet College provided support. Attendees included not only TDI agents but also administrators from the White House Office of Telecommunications Management, Chesapeake and Potomac Telephone Company, and a product manager from AT&T. The conference marked one of the first times that representatives from public and private sectors assembled on the national level to address the telecommunications needs of deaf people.

The importance of activism emerged most clearly at this conference. Two papers showed the participants how federal and state laws might be influenced by pressuring legislators. Saks gave the first paper, discussing commercial telephone answering services and problems in serving deaf subscribers. He also included copies of his correspondence with the Internal Revenue Service, showing how he, Marsters, and Weitbrecht had convinced the government to rule the cost of the coupler deductible as a medical expense.[12]

J. Thomas Rule presented a paper entitled "How the Development of a Modem for the Deaf Has Affected the Course of Deaf People's Lives." He described how a small network of deaf people in Maryland had used the TTY for activism. Through the TTY, these activists had encouraged other deaf people to write letters of support to legislators in order to pass a law to provide deaf people with sign language interpreters in court. Rule's paper portended how deaf people would become more influential in getting legislation enacted in the next decade. He summarized, "Deaf people finally have a powerful tool at their disposal to accomplish many things they could not do when there were no means to talk over the telephone. In the future, the deaf people will be more vocal about affairs around them."[13]

Marsters' paper was on the present and future status of radio teletypewriter communication for deaf people. He discussed his collaboration with the U.S. Weather Bureau in Los Angeles to establish a creative FM radio broadcast weather service for deaf people. Each day, weather forecasts were transmitted over the radio waves using the same audible tones that carry the messages over phone lines (the Baudot code). Deaf people in the Los Angeles area purchased inexpensive Radio Shack FM 4-inch "Weather radios" costing ten dollars and placed them on top of acoustic modem cradles to receive the broadcasts. Instead of "listening" to the tones from a telephone handset, the modem cradle picked up the tones from the radio.[14]

Ironically, Marsters' clever use of the TTY and modem would be rendered unnecessary by a new development that was also being demonstrated in November 1971, in Knoxville, Tennessee—closed captioned television. The federal government, through the U.S. Office of Education (now the Department of Education), would provide critical seed money to improve captioning technology over the next few years. The use of closed captions on television eventually eliminated the need for TTY news and weather services, allowing for more energy to be applied to telephone networking and relay services.

The TTY network was making some progress by this time, but the future was not clear. TDI had expanded to include sixteen local organizations, and another ten were being established. More than fifteen hundred TTYs were in use in about thirty-five states and the District of Columbia. Nearly a third of the stations were located in California and New York. Maryland accounted for about 9 percent and Missouri

8 percent. There was one TDI member in Canada and two in the Philippines. Considering there were more than thirteen million people with hearing loss in the United States, creating an extensive network was still a long way off. Also, it was still not clear whether there would be enough surplus TTYs to meet the demand. About 1,000 Model 19 TTYs and about 60,000 Model 28s remained in the Bell system. The latter machines were operating well, so it was unlikely that they would be replaced by ASCII machines in the near future. Saks met with AT&T to discuss his concern about what would happen if the supply of Baudot TTYs dried up, but no definite outcome resulted from this meeting.

Weitbrecht continued to struggle with both technical problems and with his frequently hyperbolic reaction to competition. Saks and Marsters advised him against directly accusing competitors of patent infringement or making libelous statements. APCOM's lawyer also tried to restrain Weitbrecht, and he did so by using language that an electrical scientist could understand clearly: "The signal to noise ratio in your letters to me is not always in the highest possible range and I have not thoroughly digested all of the communications you have sent me by recycling them through my receiver over and over."[15] Increasingly, it seemed that the focus of Weitbrecht's work was shifting from improving Phonetype to reviewing patent infringement details.

In February of 1972, Weitbrecht received an important commendation from part of the deaf community. The Gallaudet College Alumni Association presented him with the Laurent Clerc Award in recognition of "outstanding social contributions by a deaf person in the interest of deaf people." His invention of the modem was described as having "revolutionized the means of distance communication for deaf people and eliminated great inconveniences for them." But the notion of patent infringement and the problems, both personal and professional, posed by competition continued to annoy Weitbrecht.

As he had feared, people who purchased the ESSCO Communications ATC-2 modem and the APCOM Phonetype modem were having difficulties achieving two-way communication. As a result, the manufacturers accused one another of having poor designs. In June, Weitbrecht wrote a paper titled "Compatibility Is a Two-Way Relation" and presented a defense of his design for Phonetype. Until a solution could be found, he suggested that deaf people adjust the range finder of their teleprinters to a point between those required for each modem.

Weitbrecht was also still grappling with the modem tone loudness problem as he worked on the design of a new model, Phonetype IV. "In relation to the [tone loudness] situation, we and the competitors are stuck in the frying pan between the telephone company's inability to guarantee telephone performance with minimum loss and their demand for a limitation on transmit-level," he wrote Marsters.[16] TDI agents sometimes felt stuck in a similar frying pan as they sought to follow the organization's urging not to compare couplers or recommend particular brands.

If any events during these tense years demonstrated the importance of the work being done by the APCOM partners, Brody, and the network volunteers and agents, it was the reports of lives being saved by the TTY. TDI's newsletter for July 1972 included one such message from Jane McPherson in Kansas City: "We credit the TTY for saving the life of Mrs. Ruth Brummitt who had a heart attack while at home alone. She was able to call an interpreter."[17] Jane Miller, a deaf artist and photographer from Flushing, New York, sent letters to both Weitbrecht and Brody describing her experience. She had severe chest pains early one morning, and after telephoning a friend on her TTY, she was rushed by ambulance to the hospital. "If it weren't for TTY, you and others, I would have been dead."[18]

While deaf people recognized the obvious value of using TTYs for emergencies, convincing hearing service providers to have TTY access remained complicated and difficult. Each advance seemed to be followed by a retreat. For instance, National Association of the Deaf representative Willis Mann and Richard Clark of the Maryland Department of Vocational Rehabilitation approached the Prince George's County, Maryland, "Hotline for Youth" about expanding its network to include emergency services for deaf callers. They argued that the inability to use the telephone can isolate and terrify deaf people under the best situations, and in emergencies it can seem insurmountable. The Hotline did not become deaf accessible, however, until the Deaf Telecommunicators of Greater Washington donated a TTY and Gallaudet College loaned the organization a Phonetype modem. Within a month after the TTY Hotline was functional, sixteen deaf persons had been helped in various emergencies. After three months, the Hotline was averaging five calls a day, despite the fact that only 250 of the Metropolitan D.C. area's 5,000 deaf people had TTYs at the time. This apparent success, though, was short lived.

One problem with emergency phone lines in the days before relay service was generally available was that the telephones were often tied up with non-emergency calls, which prevented emergency calls from getting through. Those responsible for emergency services had to carefully define "emergency." To some deaf people who had no other way to relay a message to their employers, a telephone call to let the boss know they were sick appeared to be an emergency. Similarly, while canceling a doctor's appointment was not an emergency in the usual sense of this term, it was important to deaf callers. The decision to turn away non-emergency calls from deaf people was thus a painful one for personnel at emergency service centers, where hundreds of such calls came in and only a small staff of volunteers knew how to operate the TTYs. Such was the case with the Hotline for Youth in Maryland. Within a short time, the advisory committee announced that the Hotline would no longer accept answering-service-type calls and that the emergency service would end soon because of funding problems.

After a promising start, this bad news, along with more frustration from AT&T, made 1972 a disappointing year. On July 15, the *New York Times* carried an article about a touch tone signaling device developed by Bell Labs that supposedly would benefit deaf callers. Weitbrecht noted, "This contrasts with the standard Teletype keyboard which is much more convenient in operation."[19] He speculated, correctly, that the touch tone device would be no more successful than AT&T's attempts to have deaf people communicate with Morse code. Weitbrecht felt betrayed by the telephone company, thinking that by now it should have collaborated with APCOM and TDI.

The collaboration hope, however, was dashed in the fall of 1972. After two years of trying to convince AT&T to cooperate, Marsters was disappointed with their lack of response, but a chance encounter gave him a moment of optimism. While traveling on an airplane, Marsters met John R. Pierce, a professor of engineering at California Institute of Technology. Pierce offered to try to help Marsters contact others at Bell Laboratories, but he was not successful. "The net outcome [of my efforts]," he wrote Marsters in September of 1972, "is that I believe that there is *no* chance of the Bell System taking over the Applied Communications Corporation as a sort of public service. This is just to let you know that I did try to help, but that I failed."[20]

AT&T did suggest that TDI form a new committee to develop modem standards. Marsters agreed that this was a good idea, but he did not like the idea of having manufacturers on the committee. He had seen enough bickering, so he suggested that it might be better for TDI to set the standards with advice from professional engineers. TDI then collected specifications from three modem makers and consulted with the Indiana Bell Telephone Company to incorporate them into tentative standards, subject to review by AT&T engineers.

The APCOM partners and Lee Brody shared a common goal of bringing the telephone into the lives of deaf people, but their individual struggles for successful modem sales created more tensions. The fact that these principal players were themselves deaf did not lessen the friction among them. In fact, the issue of deafness became one of the battles they waged with one another. APCOM proudly boasted deaf management and deaf staff, unlike ESSCO Communications. Brody countered that APCOM contracted with businesses run by hearing people to assemble the circuitry for the modems. TDI placed a note of warning in its newsletter that although competition was the name of the game, and it was acceptable practice for manufacturers to praise and promote their own products, "this 'war' appears to have [gone] beyond the bounds of gentlemanly courtesy. . . . It is high time that such unethical and un-businesslike tactics be stopped forthwith."[21] Furthermore, TDI warned that "the spectacle of agents 'fighting' with each other makes a very bad impression on our sources of TTY's. TDI will not tolerate such squabbling. Any reports coming to us of 'fighting' . . . may result in the suspension of authorization of one or both agents."[22]

This was not the only matter of concern to TDI, particularly to Breunig, in 1972. TDI's first director was annoyed by comments he was receiving from well-intentioned people who thought the American government should assume responsibility for supplying deaf people with phone access. He wrote in the TDI Newsletter in 1972,

> There have been suggestions that the government take care of all of our problems. Teletypewriters for the Deaf, Inc., was formed by and for deaf people. The idea was that deaf persons could effectively help themselves. The growth of the network over the last four years would seem to suggest that the concept is a good one.[23]

At the same time, though, Breunig and his friends at APCOM knew that ultimately something more than the current state of volunteerism would be needed to advance telephone access. They had hoped for cooperation from private business and could not discern why AT&T was limiting its assistance to supplying discarded Baudot TTYs and Bell Laboratories continued its misplaced research emphasis on signaling devices.

By the end of 1972, the TTY network included about 2,500 machines. More than one hundred schools, vocational rehabilitation offices, and organizations serving deaf people had installed TTYs. The network's growth was all the more remarkable because TDI's home office was half of Breunig's master bedroom. Another bedroom was the shipping department. Membership records were stored in Rolodex files on a dining room table. Breunig's wife, Nancy, had been helping him since 1968 with TDI's burdensome paperwork. At this time, a new TDI board of directors was established, and the U.S. was divided into regions with representatives appointed to the board from each region. TDI also became an active member of the Council of Organizations Serving the Deaf, which was made up of advocacy groups whose purpose was to promote the welfare of deaf people.

Ever since the first issue of TDI's GA-SK Newsletter was distributed in 1970, Breunig had collected anecdotes and reports to share with the network. Many of them revealed the attitudinal problems deaf people faced in attempting to gain access to the telephone. He also reported human-interest stories to help the volunteer agents see that their work was paying off. A deaf woman in Kansas had a heart attack and managed to call for assistance at the Hearing and Speech Agency. She was rushed to the hospital in time for emergency treatment. In Washington, D.C., a suicide prevention office claimed that a life had been saved as a result of a TTY call. After a deaf man in New Jersey suffered a heart attack and died, the Telephone Pioneers of America set up an emergency service at the New Jersey State Police Station.

Deaf people quickly found that TTY emergency services and hotlines, when available, were extremely helpful. They phoned in gas leaks, reports of homes flooded from burst water pipes and damaged by windstorms, and medical situations. Personnel assisting with the answering service sometimes went beyond the call of duty. In the Washington, D.C., area, Joy York assisted a deaf person in telephoning for an ambulance. She called the doctor to meet the patient at the hospital. After finding coverage at the answering service, she personally drove to the hospital herself to interpret in sign language.

Some cities established "warmlines," which provided counseling services ranging from assistance with emergency, life-threatening situations to advice on raising a child. The first areas to provide teleministries were Philadelphia, Cincinnati, and St. Louis. Not all such services were efficient. In Washington, D.C., a deaf person called a service and received the message, "If you need spiritual guidance or an uplift from your local chaplain, you may try to ring him on number 543-9515 but you will have to try to speak in to the switchboard operator to connect you with Extension 264."[24]

The deaf community in Rochester, New York, which was expanding as the National Technical Institute for the Deaf attracted new deaf employees and many graduates remained in the city, established a creative, although temporary, telephone relay service at the Penfield Nursing Home. In return for the answering service offered by the patients, deaf people helped them with crafts. The intention was to gain some experience with general relay calls before establishing a more permanent service. Within a short time, however, the nursing home staff quickly became overwhelmed.

Saks, Marsters, and Weitbrecht followed these stories with interest. They modestly celebrated the progress of the network they had begun by saving news clippings and correspondence. Marsters stuffed them in boxes in his wine cellar. Saks meticulously categorized his clippings in file folders in his home office. Weitbrecht, living frugally with no one to share the evening hours except his dog and, when time permitted, his ham radio contacts, accumulated twenty notebooks of reports and correspondence at his APCOM office. One account told of a Glen Ridge, New Jersey, deaf couple who received the first coast-to-coast telephone call in their lives—a call from their son who announced the birth of their grandchild. In Atlanta, Georgia, the first two TTYs installed were used to help call a doctor at 4:00 A.M. to deliver another baby. In Norfolk, Virginia, an 83-year-old deaf woman living alone called the police directly when she discovered a fire started by a prankster. In Cape Cod, Massachusetts, a deaf couple was fortunate in acquiring the services of an off-duty police officer to help them move furniture into their new home. When the day was over, the officer turned down the check offered him. He felt it was his duty to report them as "bookies." The deaf couple explained to him that the large, heavy TTY in their living room was their telephone—it was not meant for taking bets on horse races. These reports also revealed how the telephone was both invading and protecting the privacy of deaf people. One deaf man had found a TTY message in the wastebasket and started divorce proceedings. The TTY printout had revealed a torrid conversation between his wife and her lover.

Despite problems, the first decade of TTY technology saw tremendous social change. The TTY opened doors for economic development. It linked deaf persons to the society of hearing people. While some telephone calls substituted for traveling, others increased the number of face-to-face encounters through new business and social interactions. Opportunities to communicate with loved ones improved, but so did those that introduced conflict. Both collaboration and competition found new avenues through the telephone.

9

FOREIGN AFFAIRS

Weitbrecht, Marsters, and Saks envisioned a time when telephone access for deaf people would be worldwide. Marsters had taken the first steps when he had introduced the modem and TTY to England and continental Europe on a tour with his family in 1966. By the early 1970s, more efforts were underway, and they highlighted many of the same issues of technological compatibility, attitudinal resistance, and hearing dominance that were prominent in the United States. Government ownership of telephone systems introduced another complicating factor.

Richard Murphy, a deaf engineer in England, faced a difficult challenge in 1971 when he attempted to start a TTY network. Telephone use in the United Kingdom was controlled by the government, specifically by the Post Office. The United Kingdom's Post Office was not receptive to the model developed by APCOM. The Post Office, like AT&T in the United States, preferred to supply its own limited teletypewriter exchange services and rental equipment. In December 1971, Murphy wrote to Marsters: "It looks like the end of our efforts to start the Phonetype system in this country and needless to say I am extremely disappointed about it."[1] Marsters wrote back in his typically crafty, intelligent style that he wished deaf people themselves would take things on and get them moving. He encouraged Murphy to

> write a member of Parliament who has a hearing impairment and who appreciates the importance of deaf people being able to use the telephone. . . . Perhaps you may be able to stir up repeated storms and realize such publicity that will encourage the Post Office to give it a try? . . . I do know a dynamic individual like you can do much to ex-

130

pedite the realization of a dream. Most of the good things in life have happened as the result of much hard work by just a few individuals.[2]

Initially, APCOM hoped that strategically placing some Phonetype modems in the United Kingdom might be a ticket to more sales. Perhaps someone there might take the lead in developing a network. Murphy's experiences with the Post Office only confirmed their suspicion that another long battle would be needed:

> I am afraid it will be most difficult for me or anyone else to start the Phonetype system here. . . . Messrs Creeds [TTY manufacturer] have completely withdrawn all help in this matter. They will not supply me with used teleprinters or technical aid—They have even taken back the two teleprinters and equipment which they gave me. . . . The P.O. authorities can do much as they like—in fact they are a law unto themselves.[3]

The APCOM partners would have to put more direct effort into the British situation, but not before they had encouraged developments in Canada.

Saks was an evangelist in the international effort. Unlike the moody Weithrecht, he stayed out of any public display of emotions in the modem war in 1972, instead focusing on business. In March of 1972, he and his wife, Jean, took a Model 31 TTY to Vancouver, British Columbia, and began a demonstration tour. They stopped at various cities all the way to Montreal. At one point, Saks called Breunig and said, I FEEL LIKE A MINISTER ON A CIRCUIT SPREADING THE GOSPEL OF THE PHONETYPE AND I FEEL I DID WELL.[4]

During the demonstration trip, the Model 31 performed very well and handled the Trimline and automatic telephones, but Saks had trouble with the General Telephone units. He had to unscrew the microphone of the telephone handset mouthpiece and drop it onto the pickup coil of the Phonetype cradle.

Marsters followed through on Saks's initiative with his usual energy. He traveled to Edmonton and encouraged Canadians to consider establishing a hotline answering service. He met with Canadian Pacific and Canadian National Railroad officials to arrange donations of

surplus TTYs and with the Alberta Government Telephone Company
to discuss modifications of Phonetype for the Canadian network. The
British Columbia Telephone Company and other companies were
then contacted for discarded TTYs. By December of 1972 there were
more than 200 TTYs in the Canadian network. It was time to look at
Europe again.

Europe hardly seemed ready for the TTY in 1972, however. Inefficient telephone systems and government resistance presented barriers everywhere. In 1971, APCOM had sent two Phonetypes to Sweden. Even though the Swedish telephone company appeared resistant
to it, the Swedish Association for the Deaf was planning to demonstrate
the TTY at its anniversary celebration. Phonetypes had also been
loaned to a man in Switzerland for a demonstration to deaf leaders in
that country, but nothing came of it. Simon Carmel, a deaf physicist
at the National Bureau of Standards in Washington, D.C., encouraged
APCOM to demonstrate the TTY at the World Federation of the Deaf
Congress in Paris in 1972. He hoped that a TTY demonstration would
be a catalyst for advocacy by deaf people in other countries. Marsters
responded that the French telephone system was a "tragicomedy" and
not worth the effort at that time.

Marsters was right about France. Telephones were still being ra-
tioned there in 1972, and only 15 percent of French families had telephones. Installation of a home phone cost $115 after approval by a
ration board. There were at least seven different types of public pay
phones, some of them in cafes, but most in post offices. What's more,
two-thirds of the busiest Paris exchanges were still using the rotary system of the 1930s. Telephones in the post offices were not available in
the evenings, Saturday afternoons, all day Sundays, and holidays. Outside the post offices, there were only seventy-eight public telephone
booths for three million citizens in Paris. London, in comparison, had
nearly 11,000 telephones on its streets. Many of the phones in France
did not even work, and the government minister in charge of the telephone system was frantically attempting to improve the efficiency of
the system.

It was against this background of uncertainty about telecommunications in Europe that Andrew Saks's hearing daughter, Andrea J. Saks, took up the challenge of bringing the telephone to deaf people in England. If ever there was a gene for telephone access, this young woman had inherited it. After leaving home for college a few years earlier, she had gotten to know her parents better—through the TTY.

> All those arguments we used to have. The eyes that closed against my silent screams. With the Phonetype I was forced to listen to what my father and mother had to say and they in turn to me. The self-awakening was enormous. . . . We became more like a normal family. . . . in the sense that though all of us, deaf or hearing, meet the same kinds of problems, communication difficulties can intensify these problems.[5]

In October 1972, Andrea Saks voyaged to London with two Phonetypes. She had a great deal of hope that she might establish an APCOM office in London and support deaf people there; however, her father and his partners had not placed any real expectations on her. The project she began took on a life of its own after she arrived. Her mother had attended the Mary Hare School for the Deaf in England and had many friends there who introduced Andrea to other deaf people. Shortly after her arrival, she attended a charity dinner for the Royal National Institute for the Deaf (RNID) at the Rembrandt Hotel in London. She arrived in time to hear the speaker, RNID chairman Rodger Sydenham, describe how deaf people in the United States were using the telephone and that the RNID was looking into it. Telephones were still far off for the British deaf community, he explained. Saks spoke up, explaining that she had just brought some modems with her. Curious about the American visitor, Sydenham invited her to the microphone to explain to the audience what she was hoping to do in the United Kingdom.

Sydenham and Anthony Burton Brown, editor of the Royal National Institute for the Deaf's magazine, became very supportive of Saks's efforts. But to get started, she needed Weitbrecht's technical expertise. She met with the Post Office, the telecommunications decision-makers, and they called Weitbrecht from London to see if he might be interested in coming to test the equipment with them. Weitbrecht was so delighted that before the phone call was over he was planning the trip in his mind.

The venture in England seemed to start off very well. Weitbrecht and Saks sparked interest among Post Office engineers to support the deaf telephone project. But at the same time, he and Saks's British contact, Andrew Kenyon, were dismayed over the huge number of TTYs rusting away in stockpiles. Weitbrecht poked around the scrapyards and asked for pieces, momentarily having reservations about the task ahead. Many of the older TTYs were being sold to poorer countries. Perhaps it was the enthusiasm of the young Andrea Saks and Andrew Kenyon, who was deaf, that made Weitbrecht slightly more confident as the days passed by.

Kenyon worked for the Post Office and belonged to Breakthrough Trust, a charitable organization of deaf and hearing people that assisted families with deaf members in finding appropriate services, organizing social events, and generally working toward improving the quality of living for British deaf persons. Before Saks arrived, he had tried to get the Post Office to look at the Phonetype modem and the TTY network in the U.S. His effort failed, so he was enthusiastic about Saks and Weitbrecht coming to London.

Weitbrecht submitted his first design for a British version of Phonetype by November 1972. Although policy was adjusted to allow acoustically coupled modems (which previously were not allowed to remain indefinitely at one phone number), the Post Office required that it be notified of every Phonetype sold in England. Weitbrecht was reminded of the restrictions AT&T had placed on deaf people four years earlier back home in the United States. The Post Office permitted the modems to be sold only to deaf individuals or their relatives. No small businesses were allowed to compete with the telephone exchange services already established. In December, the United Kingdom granted permission to APCOM to begin marketing Phonetype. However, the modems could not be sold for six months. Permission was granted only for five experimental stations in order that a study could be made of how well they worked and whether a reliable source of teleprinters could be found.

When an article appeared on December 13, 1972, in the *Daily Express* about Andrea Saks's pioneering efforts to bring the telephone to deaf people, she received encouraging letters. Philip Timms from Sheffield offered her the support of the South Yorkshire Hard of Hearing Young Peoples Club, which was affiliated with the British Association of the Hard of Hearing. "I admire your attitude in trying to convince our Post Office to have the machine available to the deaf. You've a fight on your hands."[6]

Saks viewed her work in England as a "moral issue." She wrote to Marsters of her strong commitment to her work: "I truly do believe in the basic right that all human beings have . . . to access of public communications systems and the scope of each should exceed the material gain. . . . In other words, by making Phonetype a strong factor in the lives of the English deaf we will be strengthening the position of the

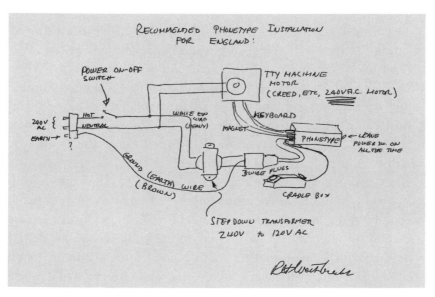

The "English Phonetype" developed in 1972 by Robert H. Weitbrecht was powered by 240 volts and used special plugs to fit British electrical outlets. Courtesy of James C. Marsters.

deaf in the entire world, that they may be able to communicate internationally."[7] She communicated this drive to her British friends as well—"To see my parents now and remember what it was like for them when I was running the show, is probably the most rewarding experience I have had. I have no words really to express my feelings. You see, Phonetype allowed them to take their own lives into their own hands."[8] In another letter she explained why she was involved in this APCOM project.

> I was the first born, and, although I have a hearing brother, I became the ringmaster of my family's communication with the hearing world. At one point you are a child and at a second point you are an adult. Within a split second you must be able to be one or the other or both at the same time. Phonetype freed me and enabled me to find myself. It freed my Mother from having to ask me to make a phone call for her which should have been private and confidential. It freed my Father to be able to conduct his business privately. It allowed us to have a normal parent/child relationship.[9]

Another important deaf person who worked with Saks in implementing the British TTY network was Michael King-Beer, an information scientist employed at the Scott Bader chemical firm. When King-Beer learned about the Breakthrough Trust, whose charitable support provided assistance to the infant network, he teamed with Arthur Cole to pioneer the "Finecom TTY System," a combination of a Type 7E Creed TTY and the British version of the Phonetype acoustic coupler developed by Weitbrecht.

By Christmas 1972, Weitbrecht was homesick. He wrote to Marsters that he had done all he wanted to do in England. "I am thinking much about my mother—she will likely be alone on Christmas Day with most of her relatives away."[10] Weitbrecht also missed his dog. It was so cold in London that he spent a great deal of time gathered around an electric heater with Andrea, her secretary, and a four-week-old puppy. He also had trouble accepting changing hairstyles. "There seem to be a lot of long hairs here—I feel out of place. . . . However, I like the puppy."[11]

Marsters believed that they needed Weitbrecht's technical expertise in England a while longer. But he also needed to warn Weitbrecht that he should proceed with tact. "I think it is most urgent that you take Andrea's advice as far as politics and business are concerned because England is so rigid and staid, something that you have to get used to. After some length, Andrea has gotten the Post Office in the right frame of mind, and you don't want to upset the applecart."[12] Weitbrecht tucked that letter into his pocket and walked to his favorite spot in the British Museum. As he had done before, he sat down and daydreamed in front of a large portrait of Sir Isaac Newton. Andrea Saks knew where to find her absent-minded friend. She was struck by the uncanny resemblance between the eccentric deaf physicist with his disheveled hair and starry eyes and the great genius in the picture on the wall.

Weitbrecht's experience in England gave him a new perspective on a recurrent theme. It was his observation that deaf people in England were less in control of their own destiny than were deaf Americans.

> There seems to be plenty of charity organizations being run by hearing people to help the deaf. . . . RNID [Royal National Institute for the Deaf] is run by hearing people—such is my impression. Will be interesting if Teleprinters for the Deaf gets started here in England. . . . It appears that there are many English deaf people who move around in the hearing world. These people will have to be educated to face the reality of the situation—they all are frustrated at not being able to use the telephone to communicate with their hearing friends. . . . It will be quite interesting if the Government [Post Office] decides to step in to force a uniform system for all deaf telephone users. Whether or not that will come to pass remains to be seen.[13]

10

REVOLUTIONS

The first decade of TTY development was marked by incremental progress against great odds. By 1973, only a few thousand TTYs were in use for the estimated 13 million Americans with hearing loss. Many deaf people were still unaccustomed to using the telephone almost a decade after the development of the Phonetype modem. People deafened in old age were often unaware of the technology, and telephone companies made little effort to reach them. There were still some states in which no deaf person had access to a TTY. Weitbrecht's modem was a major breakthrough, but refining it had taken time. Developing a reliable source of teletypewriters had also been a painfully slow process. Changing attitudes of deaf people, accustomed to years without direct phone access, and hearing people, accustomed to decades of ignoring deaf people's telecommunication needs, was a huge challenge. TTY distribution, competently administered by TDI, was still cumbersome, labor intensive, and time consuming. For years, progress had been measured in tiny, halting steps, but in the early 1970s the pace of change suddenly quickened in both the United States and abroad.

On April 1, 1973, Andrea Saks's efforts finally bore fruit when two deaf persons made the first TTY call in the United Kingdom. They used Phonetype modems in Solihull (near Birmingham) and Bromley, in Kent. The historic British call was made nearly a decade after the first American TTY call. The biggest impediment to English progress was not technology but attitude. To begin with, deaf people in England needed to verify their deafness in order to be allowed to use a telephone device. Saks wrote to Marsters that although the Post Office was beginning to recognize that the TTY was an appropriate "deaf

138

person's telephone," she had to fight for the right of hearing relatives, friends, and professionals working with deaf people to have access to the equipment. At first, the Post Office—the government agency responsible for administering telecommunications services—even objected to placing a TTY in the Royal National Institute for the Deaf building. Officials argued that big businesses should follow normal Post Office policy and use Telex (similar to TWX in the U.S.) for telecommunications data.

Transatlantic TTY calls presented another problem altogether, one that would take a long time to overcome. Saks could telephone her parents in California from England, even though the Post Office frowned on such calls. She did this by direct dialing. From the United States, however, calls to England had to go through an operator, and the FCC prohibited international TTY calls originating from the United States.

The FCC's position was related to the sending of business data, the original use of teletypewriters. At that time, the federal agency was evaluating which common carriers would share the lucrative market for sending data across the voice telephone network. Among those with stakes in the decision were the giant corporations Western Union International, ITT World Communications, RCA Global Communications, and AT&T, as well as a smaller company, Tropical Radio and Telegraph, which had a license to communicate from its ships carrying bananas. Saks convinced the comparable body in England—the Post Office—to inform the FCC of its support for her request to permit deaf people to use TTYs. She explained to officials of the common carriers that TTY calls were not just tones and signals—the calls represented communication by human beings who were alive at that moment in time. But in the minds of most U.S. telecommunications service providers, the use of a teletypewriter implied the sending of business data. As soon as the word "teletypewriter" was mentioned, an operator would deny the telephone call because of the regulations.

Saks approached this limitation on international phone access through the American political system. She placed person-to-person calls to Senators Hubert Humphrey and Alan Cranston, Congressmen

Ronald V. Dellums and John Burton, and executives of the FCC and the common carriers in an effort to persuade them. In a letter to Dellums, she explained that she had three members of Parliament willing to sponsor a TTY demonstration call. In arguing for the transatlantic call, she wrote in July of 1974 that deaf people "are beginning to play an important part in the future of the world, and their enthusiasm will be at a very high level as they have been so restricted before. . . . they deserve to have some of the telephone services that are provided for the hearing."[1]

With her deaf parents far away in California, Saks saw the need for transatlantic telephone access on a personal level, so she responded to similar concerns. One deaf woman had written to her in January for help. Her daughter was expecting her first baby, and she was married to a serviceman who was stationed in Ipswich. The woman asked Saks to make arrangements with someone in England to call them via Phonetype to give her the news that the baby had arrived safely.

But the Watergate hearings preoccupied the U.S. government representatives and Saks made little progress with her phone calls and correspondence. "I will keep on trucking and pestering till they give up and give me my way entirely!" she wrote to Andrew Kenyon.[2] Back in the United States, Marsters wrote to the FCC, explaining that it would be extremely embarrassing to have a long-awaited public demonstration between members of the British Parliament and agencies for the handicapped in the United States blocked by an overseas operator in New York. Andrea Saks also wrote to the FCC. She addressed her letter to A. C. Roseman, Chief of the International Satellite Communications Division:

> It seems rather ludicrous to me that each respective country will allow the communication of this kind of data over the public telephone network, but the U.S. carriers refuse to give permission to allow deaf individuals to use the public telephone network for communication across the Atlantic, when no other alternative is available.[3]

Saks, Marsters, and the others would have to wait until 1975 for more progress on this front, however.

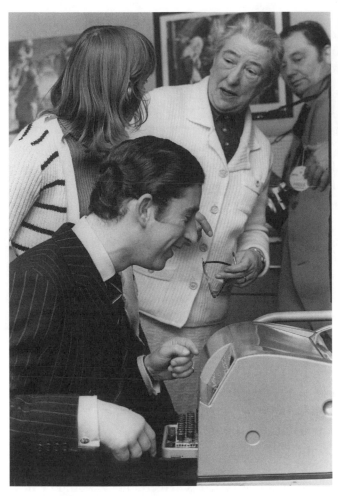

During a visit to the Maud Maxfield School in England, Prince Charles made a call on the TTY. This stop was part of a tour to learn more about the education of children with disabilities. Courtesy of Sheffield Newspapers.

A new generation of TTY technology was emerging while the struggle for international access was going on. Marsters had seen an advertisement for a light-emitting diode (LED) wristwatch. He wrote to Monsanto Company in St. Louis to see if they were interested in manufacturing an electronic telecommunications device with LED displays.

> It has been my dream to see two shirt pocket calculator keyboards put together to make a simple three row keyboard; couple this with a row

of LEDs . . . in a sort of Times Square display moving the letters from right to the left. Our PHONETYPE coupler could be reduced to a chip. This would make an excellent portable unit. . . . Would you be interested in developing such a unit?[4]

But before long, several other companies were marketing new, lightweight electronic telephone devices. Each had severe limitations. RMS Industries' TV Phone used a cathode ray tube display. Hal Communications' RTTY Visual Display System was not really lightweight, considering that it consisted of four pieces, together weighing approximately 30 pounds. The ESSCO Scan-a-Type used a Burrough's panel display, and its production was stopped after a short time because of design problems. All three were limited by their designs and expense, but they indicated what the future had in store.

Weitbrecht, Marsters, and Saks recognized that the convenience of these new desktop models, if made more maintenance free, portended the eventual demise of the behemoth TTYs, but they hoped that the price of the electronic display devices would come down so that more deaf people could have access to the telephone. The APCOM partners knew that this would not happen overnight, however. Like the transistor during its early years of production, the cost of electronic display devices would remain high for years. Mass production would be the key, and the partners thought it might be another decade before sufficient demand would exist for manufacturers to produce reasonably priced TTYs.

Political change, as much as progress in technology, would be necessary to lead to mass production of the electronic display devices. Since 1964, the goodwill of many people had served a role in giving the telecommunications movement impetus, but the volunteerism inherent in efforts to rebuild TTYs, provide relay services, or distribute machines through TDI was not enough to give deaf people equitable services and products. It only got them used Baudot machines. The telephone industry failed to respond adequately to the concerns of deaf consumers, and the IRS rulings on tax deductions did little to provide equity. It was high time for government to intervene.

The form government intervention might take, though, was important to the APCOM partners, who all believed in the critical importance of deaf people acting for themselves. Marsters articulated this feeling in a letter to his deaf friend Joe Wiedenmayer in the spring of 1973. "It frustrates me no-end to find so much money being spent by the government and by investors going up dead ends to help the handicapped," he wrote, "when the handicapped themselves should be the ones employed to determine what is practical and what is not, as well as do their own research and development."[5] The Rehabilitation Act of 1973 was a major turning point in the TTY story.

Called the "Bill of Rights" for people with disabilities, the Rehabilitation Act of 1973, particularly Title V, was similar to earlier federal statutes prohibiting discrimination on the basis of sex, race, and religion. But the 1973 Act shifted the focus of "rehabilitation" from a narrow concentration on federal funding for job training and economic opportunity to a more broadly conceived idea of social rights. The new legislation recognized the legal right of people with disabilities to equal treatment in the federal government and any services and programs that receive federal funding. The key for the telephone access movement provided by the 1973 Act was the specific inclusion of telecommunications as a rehabilitation service for persons who require use of the telephone to become employable. The 1973 law thus provided a stage for complaints to be filed and lawsuits to be pursued, particularly in later years. Subsequent amendments would expand deaf people's communication access significantly.[6] Regulations were developing slowly, however, and in 1973 deaf people did not yet have a cadre of lawyers who were knowledgeable about deafness and either deaf themselves or sympathetic to deaf people's needs.

The Rehabilitation Act of 1973 can also be seen as a marker of change that was occurring in less dramatic ways. TTYs were being installed in department stores, transportation offices, and other community organizations and services. The National Bureau of Standards joined a growing list of government agencies that had purchased TTYs, including the Government Printing Office, the Department of Health, Education and Welfare, and the Internal Revenue Service. In Washington,

D.C., the deaf librarian Alice Hagemeyer spearheaded accessible library services. In the Philadelphia area, Ralph Harwood established a radio news service so that 600 families could receive news on their TTYs.[7]

Volunteerism, though it would never be a complete solution, remained strong. Representatives of the Telephone Pioneers of America in Trenton, New Jersey; Albuquerque, New Mexico; Manhattan, New York; and Vancouver, British Columbia, helped install emergency TTY services at police headquarters.[8] Community service organizations were also helpful. Originally contacted by Marsters, Sertoma, a civic service organization with a history of supporting speech and hearing programs, provided hundreds of TTYs to agencies and individuals. An agency of the United Methodist Church, Contact, offered TTY relay services for deaf people in many cities. Numerous telephone ministries were ecumenically sponsored and made up of volunteers with expertise in counseling with regard to drugs, alcohol, sexual problems, and family and marriage issues.

Collaboration among public and private sectors also defined the growing efforts to provide telephone access. When St. Louis received a shipment of 500 TTYs from Western Union, Eli Lilly and Company provided funds to deliver the machines, and Monsanto Chemical Company offered storage space. At this time, the National Guard was in search of a community assistance project, so the 218th Electronics Installation Squadron took on the assignment to help recondition TTYs. This project alone involved a telecommunications group, a major corporation, a private business, and the National Guard.

Yet with all of this goodwill, the acquisition of a TTY in a home remained a drama in its own right. In Eau Claire, Wisconsin, when Helen Rizzi bought a TTY, the local newspaper carried a story about it. But when she asked the telephone company to list her number in the telephone book with "TTY" next to her name, they refused, thinking it was some sort of competition for the telephone. In the central Illinois area, when Mickey and Jeannie Jones requested assistance from the telephone company in Champaign to repair their TTY, the officials were not familiar with the system used by deaf people. Jones explained to the company how it worked, but the technician sent to

their home was not informed by his supervisor, and he refused to provide service. There was an angry exchange and the technician left. To
make matters worse, the telephone company billed the Joneses for the
service time, even though no service was performed. A short time later,
the University of Illinois Security Office questioned the Joneses on
whether the TTY in their campus apartment was stolen property.
Since Model 32 TTYs were being used in the campus computer laboratories, the telephone company assumed that the Joneses had taken
one of the machines from the campus facilities and alerted the university. After several more conversations, the telephone company admitted what it had done was inappropriate and offered to help repair the
Jones's TTY.

By 1974, the microminiaturization of electric circuits, a process
rapidly becoming ubiquitous, led to the development of new modems
with self-contained displays for reading text. New MAGSAT and MCM
devices, for instance, used light-emitting diodes. As a caller typed on
the keyboard or received signals from the conversant on the other end,
the text moved from right to left on a display measuring only a few
inches long. Both these machines weighed only a few pounds. Their
small size was important. They allowed deaf people to use telephone
booths. An MCM or MAGSAT could fit inside a purse or briefcase
and gave deaf people easier access to the phone while traveling.[9]
Weitbrecht had assisted several of the engineers who developed
these new products. One concern he recognized was that the electronic equipment had no paper printouts, unlike the older mechanical
teleprinters. Deaf people had become accustomed to the opportunity
to look back at the text during a telephone conversation. Since the
typed messages on MAGSAT and MCM vanished as they moved off
the left side of the electronic display, there was no way to recall what
had been typed earlier. The eventual addition of an optional printout
would solve this problem, creating a permanent record of the conversation when desired.[10]
The second generation of telephone terminals for deaf people met
only two of the criteria Weitbrecht, Marsters, and Saks had established

TTYS gradually became smaller and portable. *(Left)* RMS Industries' TV Phone displayed the text on an ordinary television screen. Manufactured in 1972, it weighed about eight pounds and rented for about $19.00 per month. *(Center)* The Micon Industries' Manual Communication Module (the MCM Communication System) came on the market in 1973, and it was one of the first lightweight, portable TTYs. The device used a single-line of red light-emitting diodes that produced an evanescent display. From the collection of I. Lee Brody, courtesy of NY-NJ Phone-TTY, Inc. Photograph by George Potanovic, Jr./Sun Studios. *(Right)* The MAGSAT TTY also used light-emitting diodes and weighed less than five pounds. Courtesy of Telecommunications for the Deaf, Inc.

ten years earlier. The solid-state devices were available and portable, but they were not yet affordable. The electromechanical TTY and modem cost between $200 and $250 by 1974. The new electronic devices cost between $600 and $1,000, much too steep for the average deaf worker, who was earning about 70 percent of the income brought home by the average hearing worker.

A 1974 Stanford Research Institute survey showed how serious the cost issue was. First, the survey indicated that only 18 percent of deaf respondents were using *any* telephone device. Second, 62 percent of the respondents were unwilling to pay more than $200 to own a TTY, and 91 percent were unwilling to lease a device for more than $10 per month.[11] This resistance was one reason why there were few subscribers to the RMS Industries TV Phone, which had a monthly rental rate. It just was not economically sound to rent a TTY and then pay regular telephone bills and long-distance fees, especially when there were relatively few deaf people to call, and relay services for telephoning hearing people were highly inefficient.

In the midst of these exciting times, Weitbrecht's personal difficulties and professional frustrations seemed to increase. His mother, one of the few people with whom he was very close, died. He read that the Bell system's new Trimline telephone, a compact model with the dial on the handset, emitted signals too weak for hearing aid users. More than 1.5 million persons with hearing aids would not be able to use these telephones or the new receivers due for installation in pay phones. Once again, hearing people created barriers for deaf people.

In an April 1974 essay on access and equity issues Weitbrecht revealed his unhappiness with a world dominated by the interests of those with hearing. He emphasized that during the first decade of TTY development no government assistance was accepted. The network of deaf telephone users, he wrote, "is really of, by, and for the deaf people." He estimated that the telephone-teletypewriter network had "brought millions of dollars in added revenues to the operating telephone companies, as a continuing result of new telephone installations and long-distance telephone-call charges." With 7,000 teleprinters in use, costing an average of $250 (including modems), the deaf community had an investment of about $1,175,000 in the old equipment. He worried that now there might be a mandated replacement of the older teleprinters by the newer electronic devices, and he stressed that deaf people as a group wanted to be "allowed to evolve their own communication needs. . . . Deaf people have every right to be consulted as to their own needs—not to be dictated to by the hearing people. . . . Ideally, the deaf people will be happy to participate in various decisions as may affect their future, provided such decisions are reached with full understanding on the parts of both deaf and hearing peoples concerned."[12]

Weitbrecht was financially strapped, too. Some of the very people he was helping should have been paying him royalties, at least from his perspective. He was making little money on the Phonetype patent. APCOM paid him token royalties, with most of the revenue from sales going to research and development, especially to improve the Phonetype modem. Weitbrecht identified with Edward E. Kleinschmidt, who had evolved the teleprinter back in the 1920s and who had acquired 117 patents by 1966 when he was 91 years old. Like Kleinschmidt, Weitbrecht was persistent and creative but never seemed to be able to make money, as others benefited from his patents. In the spring of 1974, Weitbrecht lapsed into melancholy over his days as an astronomer. He wrote to Marsters, "I was a pretty lonely star gazer;" he metaphorically referred to himself as an "intergalactic tramp, moving in the void between the galaxy of the hearing people and the galaxy of the deaf people."[13]

Much of Weitbrecht's time continued to be spent in ultimately fruitless correspondence with his lawyer in investigating possible patent infringements. He believed that his patent was being violated by those who had manufactured competing modems as well as by the newer electronic devices. His lawyer was an electrical expert, and they exchanged numerous long letters covering such topics as "frequency discriminators," "phase locked loops," and other components of the circuits. The question of patent infringement could ultimately be resolved only in court, however, and without money to pay lawyers, APCOM was advised to give up.

At least Weitbrecht was receiving recognition within the American deaf community. In 1974, he received an honorary doctoral degree from Gallaudet College for his "extensive research and service to the deaf community as a physicist and electronic scientist." Marsters had nominated him and encouraged many leaders in the deaf community to write letters of support for this award. This was a particularly happy event for Weitbrecht, and one filled with a moment of adventure with telephone access as well. While in Washington, D.C., to receive the award, he had great difficulty hailing a taxi to take him to the ceremony. Stubbornly independent, he had attempted for some time to wave down a cab on the streets. Finally, he asked the hotel manager to telephone the Gallaudet Alumni Office to leave a message that he would be late. A mad scramble ensued, and the hotel manager personally drove Weitbrecht there. The deaf physicist took his place just in time, to everyone's relief, next to two nationally known deaf leaders, Boyce R. Williams and Merv Garretson.

Even on this important weekend, Weitbrecht found time to help others with their telephone problems before returning home. Stopping at the Captioned Films Bureau, he repaired a "drunken" line-feed linkage in a Model 32 TTY. Although very appreciative of the award, Weitbrecht felt that his partner should have been similarly recognized. After returning to California, Weitbrecht wrote a letter of appreciation to Gallaudet President Edward C. Merrill, Jr.: "Someday I hope to see Dr. Marsters similarly honored; without him, I would have never entered the field, and there would have been continued difficulties in getting telephones to work for the deaf."[14]

It was at this time that Weitbrecht was called to assist TDI with an important project. Despite the promise of the new electronic terminals, their expense meant that thousands of older devices would continue to be used. In 1974 nine out of ten deaf people stayed with the bulky electromechanical teleprinters connected to modems. Therefore, TDI needed a training manual to guide people in reconditioning and maintaining them. Breunig had acquired funding from the Eli Lilly Endowment in Indianapolis, and Paul Taylor invited Weitbrecht to St. Louis to help him and six deaf friends. For more than 100 hours, they tore apart various TTY models, photographed the parts, and wrote and edited a 300-page manual entitled *Teletypewriters Made Easy!* The "Red Book" as it became known in the deaf community, was printed just in time for TDI's First Convention of the International Teletypewriters for the Deaf, Inc., in Chicago in June 1974.[15]

In Chicago, TDI's president, Latham Breunig, a key player in the TTY story, fell victim to the communications schism within the deaf community. The politics of diffusing telephone technology had accelerated quickly within the new environment of technological and social revolution. In the process, some of the spirit of collaboration among TDI's agents and officers, who had been drawn from both the signing and oral traditions within the deaf community, was lost. Many TDI agents discovered that they were unable to communicate with Breunig in person. The retired chemist and statistician was a staunch oralist who did not support the use of sign language. At the TDI business meeting in Chicago, he needed the assistance of a sign language interpreter. Many in TDI now believed that his communication preferences made him an inappropriate liaison for the deaf community, and Al Pimentel, a skilled user of American Sign Language, was voted TDI president. Breunig was bitter over this decision: "For years the manual-ists have been saying 'let's get together' for the good of all deaf people," he wrote to Andrea Saks; "TDI has done just this, getting them to-gether. But are they satisfied? No, not unless this togetherness is on their own terms." In administering TDI, he had never asked a prospec-tive agent about personal forms of communication. "I guess there were some disturbed by this 'hang-up' on my part and this played into the hands of my enemies."[16]

Outside of the politics of communication, this first TDI conven-tion was successful and ended on a humorous note. During the ban-quet on the final evening, a raffle was held. As Weitbrecht reminisced in a letter to Andrea Saks in London, "It was very funny. . . . there was a final drawing, and Lee Brody won the Phonetype IV! So I went over and shook hands with him, congratulating him on winning such a fine product."[17]

Brody had remained active in deaf telecommunication issues. At the time of the TDI convention, he was involved in work on a Braille TTY. When he returned to New Jersey, he put the finishing touches on this new machine for people who were both deaf and blind. His terminal included a Braille Embosser that allowed a person unable to see or hear to receive a TTY message through impressions made on a peripheral punch tape. In October of 1974, Brody's device was pub-licly demonstrated. The first Braille TTY call was made between Martin Sternberg of New York University and Robert Smithdas, a deaf-blind man working at the Helen Keller National Center for the Deaf-Blind in Sands Point, New York. Following this call, Brody installed a Braille TTY at the Gallaudet College National Academy for a deaf-blind man, Arthur Roehrig, who was assistant to TDI president Al Pimentel.

Brody met with a range of emotions when he introduced the new Braille telephone device. Many deaf-blind people refused the technol-ogy, worrying about their typing abilities and what others might think of their intelligence when seeing their efforts to communicate. It was

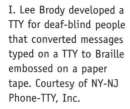

I. Lee Brody developed a TTY for deaf-blind people that converted messages typed on a TTY to Braille embossed on a paper tape. Courtesy of NY-NJ Phone-TTY, Inc.

a different world than he was used to, one of both darkness and silence, and Brody was touched by the needs of deaf-blind persons. When he visited a deaf-blind man in Cleveland, he rang the doorbell, and the floor of the entire house vibrated. Electric motors had been connected to the doorbell. During a power outage one winter, the man crawled into bed to stay warm and plugged in his electric razor. He placed it next to his leg where its vibrations would let him know the power had been restored. When he awoke from the vibrations, however, he confused them with the shaking floor, thinking that someone was at his door.

The cost for one Braille Phone-TTY station in 1974 was over $2000. But Brody did not wish to sell stations to deaf-blind people. Rather, he solicited sponsors to finance the cost of the equipment to be installed and serviced. Within two years, he had installed more than forty Braille TTYs around the country. When word spread about the Braille TTY, he received many letters. From Kansas City, Kansas, came a letter written by a neighbor of a deaf-blind woman: "I can't help her much—If she has a TTY, it would be more safer for her to talk on it instead of coming over to my house. Many times she gets lost and neighbors help her find my house."[18] Another letter came from Danville, Kentucky, and described a blind mother with six children. "Now you see that four of them are deaf and all have their TTYs in their houses. They asked me hopefully to write to you to see if you could help them by providing the TTY-Braille machine for their mother so she could communicate with her deaf children directly."[19]

Like his APCOM competitors, Brody never lost sight of the real purpose of the TTY venture. He filled his warehouse with donated TTYs, hoping to recondition them in time. During one trip to New York City, he amazed everyone by maneuvering an eighteen-wheeler through a narrow section of Wall Street in order to pick up a load of TTYs. Brody later received a prescient letter from President Ronald Reagan about his work with deaf and deaf-blind people: "To be cut off from the ability to communicate with others can be a lonely, frightening, and even life-threatening experience . . . I know their gratitude for your effort is boundless."[20]

Despite the revolution in communication access occurring on many fronts by the early 1970s, access to telephone relay services remained one of the most intractable problems. The efforts to establish a hotline in the Twin Cities area in Minnesota illustrate the problems deaf people were having all around the country. When local leaders approached Vern Heglund, a customer relations department manager of Northwestern Bell Telephone Company, to get his support, he was initially hesitant. Then he spent some time with deaf people and became aware of some of the day-to-day problems they encountered. They showed him the Linotype operation at the Minneapolis *Star-Tribune*, where about forty deaf workers were employed. They told him how they could be assisted by an answering service that would enable them to communicate with non-TTY users, especially in emergency situations. Northwestern Bell would not donate the personnel, time, and equipment to set up a relay service, but the company made arrangements with A-Alpha Answering Service in Minneapolis to handle relay communications for fifty cents per local call. On one hand, it may be argued that this represented progress. On the other hand, when one considers the accumulated cost of having to pay fifty cents per call, it seemed to deaf people that every telephone-related service they attempted to establish required great struggle and expense. "Equity" remained a distant dream.

11

BRIDGES

On May 13, 1975, Weitbrecht was delighted to participate in the first authorized transatlantic TTY call using the modem he had developed. Finally, nearly two decades after the first transatlantic telephone cable was laid and more than a decade after the Phonetype for deaf people was developed, the FCC had granted AT&T temporary authority to provide the use of its facilities for a "one-day demonstration of Phonetype." At 12:30 P.M. (7:30 A.M. in Washington, D.C.), Britain and America at last were linked in an authorized TTY call across the Atlantic.

Andrea Saks brought this about. The FCC told her that if she were able to get all of the common carriers to agree, they would grant her a one-day-only waiver while new telecommunication regulations were before the FCC for consideration. The U.S. Department of Commerce Telecommunications Equipment and Systems Exhibition at the United States Trade Center in London was planned for May. More than sixty U.S. firms would be presenting their latest state-of-the-art technology to the British market, and the FCC believed that the demonstration call between deaf people would be good publicity. Saks cleverly figured out how to get all the telecommunications carriers to agree.

She used her college training in acting to promote her cause. Saks direct dialed each common carrier representative and pretended to be a long distance operator calling from London. Once she completed the connection, she changed roles and made personal appointments with the carrier representatives because she knew that she would not be convincing over the telephone. With AT&T, the FCC, and the Post

Andrea J. Saks played a significant role in gaining permission for the first transatlantic TTY call in May 1975. On the day of the call, she held a press conference at the U.S. Trade Center Building in London. She has spent most of her adult life in pursuit of equity and compatibility in international telecommunications. Courtesy of P.I.C. PHOTOS LIMITED.

Office in England all supportive of her efforts, she was able to convince the carrier representatives to see her. Saks then flew to New York City to meet with each carrier. In this way, she personally succeeded in getting all of the approvals for the demonstration TTY call. The venue for the call came together very fast after the New York City meetings.

The demonstration call was highly publicized. During the first TTY conversation, Weitbrecht typed to Jack Ashley, a deaf Member of Parliament: **THIS IS THE UNITED STATES OF AMERICA CALLING GREAT BRITAIN.** Ashley responded,

GOOD AFTERNOON. THIS IS LONDON SENDING WARM-EST GREETINGS TO OUR FRIENDS IN THE UNITED

STATES. . . . CONGRATULATIONS ON THIS SYM-
BOL OF PROGRESS FOR DEAF PEOPLE WHICH HAS
GIVEN VAST ENCOURAGEMENT TO EVERYONE HERE.[1]

Ashley also told the press that "the people who have made this call possible to the United States deserve a vote of thanks from the deaf people throughout the world. It is a monumental achievement. What it's doing is pushing forward a frontier, and any person or organization who can accomplish that has every reason to be proud."[2] Ironically, these years saw Weitbrecht, the principal developer of the TTY modem and Ashley's partner in this historic conversation, falling by the wayside in the telecommunications revolution.

By mid-1975, the number of TTYs in the United States had reached about 10,000. With this growth, the political momentum in the deaf community increased. President Ford's signing of Public Law 94-142, the Education for All Handicapped Children Act, ensured access to free, appropriate public education and assisted states and localities with funding to deliver effective educational programming. The law also reflected general increased acceptance of diversity in American society. In 1975 the National Center for Law and the Deaf was established at Gallaudet College to help eliminate barriers for deaf people in education, employment, health care, legal services, and government programs. The staff members were experts in legal issues related to disability rights and trained in sign language. One of their first actions was to encourage AT&T officials to form a high-level management committee to review the needs of people with disabilities.

Public activism was breaking out everywhere, and protests and legislative enactments brought more visible access on a much broader scale than ever before. In South Dakota, for example, Benjamin Soukup, a deaf man, pushed to establish a statewide TTY-voice relay service using a toll-free number. This meant that deaf people there could make long-distance calls twenty-four hours a day, seven days a week, by dialing trained staff in the state-run service. In Los Angeles,

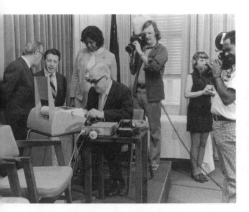 

(Left) The first authorized transatlantic TTY call took place on May 13, 1975. On the American side, Boyce R. Williams, Director of the Office of Deafness and Communicative Disorders, Rehabilitation Services Administration of the Department of Health, Education and Welfare sat at the TTY. He was joined by *(left to right)* Karl Bakke, Acting Secretary, Department of Commerce; British Ambassador Sir Peter Ramsbotham; the Honorable Caspar Weinberger, Secretary, Department of HEW; and Charlotte A. Coffield, Program Specialist, Office of Deafness and Communicative Disorders.

(Right) On the British side at the U.S. Trade Center Building in London, deaf Member of Parliament Jack Ashley typed his message to the Americans. Behind him were Michael King-Beer, a deaf pioneer with the British TTY network, and U.S. Ambassador Elliot L. Richardson. Photographs courtesy of P.I.C. PHOTOS LIMITED.

deaf people and their hearing friends were waging an intense battle for equipment access. Florian A. Caligiuri, a deaf employee of California State University, Northridge, was on the frontlines. He argued that deaf people not only paid an average of $250 for a TTY but also additional costs for upkeep, as well as extra charges for long-distance calls. With such costs, very few deaf people were able to purchase TTYs. "Simple arithmetic shows that there is only one TTY to every 530 deaf persons, or one TTY to every 4,000 with a hearing impairment," he wrote. Caligiuri emphasized that until more telephone devices were made available, deaf persons would continue to be penalized by the inaccessibility of telecommunications: "A person who is either deaf or for whom amplification equipment is of little or no value cannot pick up the phone and make a purchase at a department store. He or she cannot contact the police in an emergency. In fact, deaf people are deprived of all the services available to hearing people and which they take for granted."[3]

The Pacific Telephone Company had developed bilingual phone service for Spanish-speaking Californians, but there was no adequate service for deaf persons. Caligiuri suggested that AT&T subsidize the manufacture of telephone equipment for deaf people and distribute it the same way phones for hearing people were distributed. The Greater Los Angeles Council on the Deaf agreed and wrote to Pacific Telephone in 1975:

> It is our contention that the telephone company has an obligation to make equipment available to hearing impaired and other customers at a cost comparable to that at which ordinary telephone equipment is made available. The residential customer of the telephone company is not required to buy his or her expensive telephone. Rather he or she rents it at some low prearranged monthly charge. We contend that a deaf person should be able to rent an appropriate telecommunication unit at a comparable rate.[4]

By 1976, social and technological change had magnified deaf people's desire for full communication accessibility as they and their families saw unjust practices, inequity, and discrimination in the failure to apply appropriate technology to their communication needs. Deaf leaders spoke out. Psychologist Larry Stewart, for example, bemoaned the lack of support from telephone companies: "The patience of the silent minority is growing thin. How much longer must they wait for the freedom, justice and equality promised to all Americans?"[5] To further prepare for the battles ahead, the National Association of the Deaf established its Legal Defense Fund. With the Legal Defense Fund and the National Center for Law and the Deaf, the deaf community finally had strong legal voices.

Demonstrations against the phone companies increased. In one protest, deaf people gathered in front of the Bell Telephone Building in Pittsburgh to express their concern over the expense of phone service. They urged Bell Telephone to work on special telephone equipment at reasonable prices. In Denver, deaf demonstrators picketed outside the downtown offices of Mountain Bell, taking issue with the fees for equipment that deaf persons needed for use of the telephone. Although the Public Utilities Commission required them to pay these fees, deaf people did not even have access to directory-assistance information and could not report an emergency through the 911 system.

As the telecommunications struggle shifted toward the political arena and technology moved on to microelectronics, Weitbrecht was left further behind. He appeared unable to let old battles die, and he grew more bitter. He wrote to Lick Observatory complaining about not being asked to consult in recent modifications made to the camera that he had designed many years earlier. He wrote to the American Association for the Advancement of Science Project on the Handicapped in Science about the discrimination he had experienced earlier in his career. He also wrote to his friend George Fellendorf at the A. G. Bell Association about his being too busy to pursue his astronomical interests.

> My childhood interest in astronomy has suffered a decline. I have already dropped the subscription to one national astronomical journal, and I am thinking of terminating my life membership in one quasi-professional astronomical society . . . I should make an effort to eliminate all references and reading material which only remind me of astronomy. It bothers me greatly to know that I am unable to contribute effectively to that field. . . . It is a hard life.[6]

Competition was not making things any easier for Weitbrecht. By 1976, APCOM's sales of Phonetype had begun to slow down.

In October 1976, Weitbrecht's friend Barbara Chertok reminded him of how far away they remained from the goal of a national relay service to connect TTY calls with voice calls. An acquaintance was paying fifty dollars per month for an answering service to perform this role. Weitbrecht thought that was much too steep a fee. Many other deaf people still depended on volunteers working out of private homes to provide relay services, a solution Weitbrecht found unsatisfactory. YOU MUST REMEMBER THAT HEARING PEOPLE TEND TO FIND

IT AGGRAVATING TO HAVE TO ''HELP'' DEAF PEOPLE, he told Chertok.[7]

Marsters was adjusting to the rapidly changing political and technological conditions much better than Weitbrecht was. Joining the activism, he provided input to a task force—Services for the Handicapped—commissioned by Pacific Telephone. Particular emphasis was being placed on telephone communications. The task force report indicated that consumers with disabilities were generally unaware of current offerings and had difficulty contacting the appropriate telephone company personnel for assistance. In January 1977, Pacific Telephone reached a conclusion that had been obvious to the APCOM partners for years: Their "telephone sets and ancillary devices" permitted "at best . . . only a crude and cumbersome form of communication."[8]

In February, Pacific Telephone's Task Force met with a group of people to compare advantages of the TTY/modem system and the TV Phone, MCM, MAGSAT, and HAL communications devices for possible use in meeting the needs of deaf Californians. The group included Leo Jacobs, a leader in the deaf community, and Steve Jamison of IBM, the father of a deaf boy. There were now more than 27,000 TTYs in the country, and California was preparing to take the lead in investigating state support for an equipment access program.

Nationally, the civil rights movement for people with disabilities had stalled temporarily. Even though the 1973 Rehabilitation Act was several years old, but the regulations had yet to be implemented. In April, deaf people joined thousands of people with other disabilities in staging an angry sit-in at nine regional offices of the Department of Health, Education and Welfare to demand that the regulations be signed and enforced. Frank Bowe, the director of the American Coalition of Citizens with Disabilities (and who is deaf), was one of the protest leaders. He told a news conference on Capitol Hill that Health, Education and Welfare Secretary Joseph A. Califano, Jr., had "failed to fulfill the promise of equal rights" for people with disabilities by refusing to implement Section 504 of the 1973 Rehabilitation Act. After Califano finished a ten-minute speech over the angry shouting of the crowd, Bowe stood up and suggested that he sign the regulations now and review the rules for possible changes later. George Reed, a sixty-five-year-old blind man, told a reporter, "I marched for my civil rights as a black man in the '60s. I never thought I'd see this day come when handicapped people would rise up and demand their rights. We've been begging for a long time. Now we're demanding our rights."[9] Califano signed the regulations three weeks later, effective June 1, setting the stage for sweeping changes in employment practices and services.

The same month of the historic sit-in, Congresswoman Gladys Spellman of Maryland, who represented a substantial number of deaf constituents, introduced a bill in the U.S. House of Representatives that required Congress and federal government agencies to be accessible to deaf people by telephone. Working with the National Center for Law and the Deaf and the National Association of the Deaf, Spellman also requested special telephone rates equitable to those hearing people pay for a call to the government. By October 1977, Senator Lowell Weicker had installed two TTYs in his Bridgeport and Hartford, Connecticut, offices to serve his 180,000 deaf constituents.

Meanwhile, APCOM's troubles and dilemmas worsened. With falling sales, the company was not bringing in adequate revenues to continue all its activities, but money was still needed to support Weitbrecht's research and development efforts. In order to economize, the partners decided to discontinue support for Andrea Saks's advocacy work in London. Andrea, back in Redwood City for a while, argued bitterly but unsuccessfully against the decision. She predicted, correctly, that closing APCOM's office in London would eventually lead to serious incompatibility problems as others pushed for telecommunications terminals using different transmission codes. She then left for England, hoping to somehow find a way to undo the damage that she knew would result from the closing down of the London APCOM office.

Despite APCOM's problems, the general telecommunications movement "of, by, and for" deaf people in the United States was gaining momentum; deaf leaders were gaining confidence, and at least some companies were beginning to understand that deaf people's desire for communication devices of various kinds represented a potentially lucrative market. In 1977, for example, the exhibit hall of TDI's convention in New York was full of vendors. When a visitor from AT&T stopped by the convention, Lee Brody moderated an impromptu meeting. The AT&T official was unable to answer most of the questions, however, and Alfred Sonnenstrahl asked him why he was attending. The AT&T representative explained that he was a manager of a branch office down the street and admitted that an hour earlier he had been encouraged to stop by the convention. He had never heard of a TTY even though AT&T was making millions of dollars in revenue from deaf people by then.

The revolution was further bolstered by the White House Conference on the Handicapped that year. There were forty-four deaf delegates. Among the "social concerns" expressed at this conference was that Americans who were deaf or blind had "profited very little from the remarkable scientific and technological advances currently being made in other fields."[10] Funds allocated by the federal government for research related to the development of sensory aids remained minimal. About $220 was expended annually per cancer patient and $76 per cardiovascular patient, but only $.41 was allocated for each deaf person.

As sales continued to decrease at APCOM, the three partners discussed closing the company, but Weitbrecht's status was a major problem. Earlier, William E. Castle, the director of the National Technical Institute for the Deaf at Rochester Institute of Technology, and his wife Diane, a professor at the college, had encouraged Weitbrecht to consider teaching deaf college students. Saks suggested to Marsters that they prompt Weitbrecht to think about this again. In December, APCOM received bad news from Pacific Telephone. The telephone company was to have an operating system in place for deaf customers early in 1978, and the APCOM Phonetype had been considered for use. At the last minute, however, the requirement of automatic answering precluded Phonetype. Saks attempted to communicate with telephone company officials, but his relay service gave him trouble. He quickly sought another company. Much to his chagrin, it was extremely difficult to find another answering service that would accept deaf customers using TTYs. APCOM remained in business but its position was weakening.[11]

To ease the burden on Weitbrecht, Marsters and Saks hired Cesar Barrientos, a technician, to assist with electronics design work. But APCOM was still unable to keep up with the increasing costs and fast pace of the telecommunications revolution. With little capital for continued research and development, Marsters and Saks studied the possibility of a merger or buyout for APCOM.

Weitbrecht still had no comfortable retirement plan and no other source of income. Most of Saks's money was tied up in a trust fund, so he was unable to help Weitbrecht further. He and Marsters were happy to have helped Weitbrecht get started back in 1964, and they provided modest royalties to him over the years without any compensation to themselves. But Saks recognized Weitbrecht's lifelong inattentiveness toward his personal finances. **BEING A SCIENTIST, HE NEVER GAVE A THOUGHT ABOUT SECURITY FOR HIS OLD AGE TO HIS REGRET. HE KNOWS NOW THAT HE SHOULD HAVE THOUGHT OF PLANNING AHEAD.**[12]

Weitbrecht, now completing the design for Phonetype XI and working with Cesar Barrientos on a thermal printer, purposefully ignored all discussions of APCOM closing. In April 1978 he wrote a letter to Saks and Marsters explaining that APCOM's slow progress was related to problems in finding appropriate keyboard and display components for a new model they were planning. Saks did not want to hear any more about the business, but Weitbrecht stubbornly disregarded his partners' concerns about their own retirement planning. He made little comment about mergers and wanted to see APCOM prosper and be of continued service to deaf people.

By 1978, the federal regulatory system was beginning to understand deaf people's demands. In that year, the National Center for Law and the Deaf at Gallaudet College petitioned the FCC to begin a formal inquiry into the telecommunications needs of deaf people. The attorneys argued that the current policy of high cost and poor service constituted discrimination. A proposal entitled "Telecommunication Services for the Deaf and Hearing Impaired" (Docket 78–50), sought to establish a national forum where the communications common carriers, manufacturers of telecommunications equipment, and deaf people could interact. Issues included a "universal telecommunications device," specialized communications services, and the provision of services by telephone companies. Weitbrecht participated actively in the inquiry, drafting a lengthy response to the proposal. His letter to the FCC included a revised upward estimate of the number of TTY units used by deaf people. The FCC estimated the number at 20,000, but Weitbrecht thought there were at least another 10,000 to 15,000 in use. Weitbrecht also defended the use of the Baudot rather than ASCII communication code. His letter reflected how he was falling behind in the revolution of technologies.

Deaf people provided the FCC with input in a number of ways. TDI encouraged TTY owners to let the FCC know how their telecommunications needs could be met. A TTY News Service developed by the Learning Resources Center at the Model Secondary School for the Deaf at Gallaudet College gave a public demonstration of telephone devices used by deaf people. Calls were solicited from around the country to provide input to the inquiry. Deaf children described how they wished to communicate with their parents but could not afford the devices. In many cases, their parents could not afford them either. Some children mentioned how the TTY helped protect deaf people from crime; others emphasized the importance of being equal to hearing persons. There were also no TTYs in public phone booths, which frustrated deaf travelers.

With an audience of spectators looking on, one demonstration TTY call was placed to a supervising teacher at a school for deaf students in the West. The teacher had been called out of town, however, and the call was transferred to the office of the dean, who was not informed of its purpose. "I am a very busy person," he typed as his message was projected on the TV screen in Washington, D.C. "Have we talked enough for your demonstration . . . I do not know what purpose it serves."[13]

Deaf leaders also were making some progress with their old nemesis, AT&T. In early 1978, Alfred Sonnenstrahl and Jay Croft, a deaf minister, were invited to meet in Omaha, Nebraska, with about ten AT&T executives. Sonnenstrahl and Croft pressed AT&T to consider changes that would benefit deaf people, including the formation of operator services for deaf people, long-distance call discounts, publication of TTY directories, and employment of deaf workers. AT&T eventually responded by creating a customer relations office and hiring Sarah Jepsen, who coordinated monthly consumer panels. Within a short time, there were operator services for deaf people in New York City, Philadelphia, Oakland, and Omaha. Furthermore, as part of its planning for products and services, AT&T had conducted a study indicating that 16,650,000 people in the United States were deaf or hard of hearing. Another 8,000 people with normal hearing were unable to use the telephone because of severe speech-related disabilities. While these estimates were low, they nevertheless revealed the need for attention to these consumer groups.

Progress with Ma Bell was neither easy nor consistent, however, as Marsters found out again in July of 1978. He had connected a Wheelock relay device to his phone line in order to have a signaler lamp flash when the phone was ringing. Pacific Telephone and Telegraph, on whose communication task force Marsters served, then sent him a letter threatening to terminate his telephone service. He was told that if he did not supply the relay device manufacturer's name, equipment model number, ringer equivalence number, and other information within ten days, he would be without a telephone: "If the condition has not been corrected, and I have not received your written notification by August 9," the letter stated, "your service will be suspended."[14] To the man who first conceived the idea of a TTY network for deaf people that now brought millions of dollars to AT&T, this letter was ironic.

The legal system provided another venue for deaf activism. In April of 1978, the National Association of the Deaf Legal Defense Fund and Charles Crowe, a deaf man from Morganton, North Carolina, filed a complaint with the Architectural and Transportation Barriers Compliance Board against the Office of Civil Rights of the Department of Health, Education and Welfare. The Civil Rights office had told

Crowe "to call collect on the telephone to talk about" a civil rights case.[15] The office had no TTY, however, and Crowe was unable to call them directly. He charged that the department had not installed tele-communication devices, thus making itself inaccessible to some two million deaf people in the United States. As a result, Health, Education and Welfare installed TTYs in its headquarters and regional offices. In August 1978 deaf people in Maine filed a formal complaint with the Public Utilities Commission, charging New England Telephone and Telegraph with "unjust, unreasonable and discriminatory" rate structures and services.[16]

Legislative action on the state level also became an important tactic. In August 1978, Michigan Senator Gary S. Corbin introduced a bill claiming that the Bell Telephone system had been discriminating against deaf people who could not afford equipment. His bill required that telephone companies lease TTYs to deaf people unable to afford to buy the devices. Other states began enacting similar legislation, leading AT&T to issue a policy statement to operating companies recommending that telephone companies make no profit on such terminal equipment.

Although deaf people were making tremendous progress in the telecommunications arena, the battles had taken their toll on Weitbrecht. As 1978 came to a close, he was, at 58 years old, tired of fighting. He still saw a role for himself in APCOM's research and development program, but as the TTY movement grew rapidly, he felt tugged in many directions. The telephone companies continued to look to him to respond to their concerns about modem specifications. Other companies developing new products asked him for guidance. He remained as passionate as ever about the importance of deaf people controlling their own lives, however, and as disappointed as ever with his inability to convert his invention of the acoustic coupler into a profit-making device.

Both of these characteristics were in evidence in late 1978, when he learned about the work of his deaf friend Paul Taylor in St. Louis. Taylor and his colleagues developed the "C-Phone," a large television screen communications device. Weitbrecht telephoned Saks to discuss the new TTY: THEY HAVE REASON TO BE PROUD OF THE FACT THAT THEY WERE NOT HELPED BY HEARING PEOPLE. THAT IS WHAT I LIKE MOST ABOUT C PHONE.[17]

Saks knew that each time a new device came out, Weitbrecht became momentarily disappointed that he received no royalties for his own patent. Saks attempted to counsel his friend with a touch of humor: YOU HAVE LEFT YOUR MARK THAT WILL GO DOWN IN HISTORY. THINK OF THE POOR CHILDREN WHO WILL HAVE TO LEARN HOW TO SPELL YOUR NAME.[18]

12

THE CHANGING OF THE GUARD

The early months of 1979 were plagued with personal crises for Weitbrecht. He mailed more letters to the University of California complaining about his earlier employment experiences there. He wrote to Marsters his neighbors were complaining about the ham antenna on his property because it interfered with local radio reception. The neighbors also had problems with his newly adopted dog, Mickey, a black Irish Setter. "What bothers me now about the next door tenant . . . she is doing a beautification project right on my land . . . she planted posies and was pulling up weeds (on my land) . . . Mickey wandered over to one of the trash bags and peed on it. . . . That really drove the woman up the wall."[1] He also wrote to Barbara Chertok that he was "fed up with competing with other companies for share of the TDD market." Perhaps even more telling, he explained that "both of [my partners] have considerable independent incomes. I do not have anything to fall back upon except my small savings and income from rental of my other house. The other competitors have never paid me a red cent of royalty because of my patent."[2]

APCOM's situation was growing desperate as well. Weitbrecht refused to concede his partners' claims that APCOM was no longer a force in the TTY movement. He concocted projects—more work on thermal printers, improvements in the Phonetype design, and other efforts that produced little success. Weeks went by as Marsters and Saks, who were carrying APCOM financially, pondered ways to close APCOM and yet provide Weitbrecht with a retirement plan. Several mergers with other companies were considered, but Weitbrecht rejected them. His uncertainty about his future made him increasingly hard to work with. APCOM's marketing manager Jeanne Poremba and

168

the other staff regularly contacted Saks and Marsters for advice on how to deal with the despondent and aimless physicist.

In September of 1979, Weitbrecht finally conceded that Marsters and Saks had carried the financial burden too long and that the business was not breaking even. Marsters attempted to console him:

I CONSIDER YOU ONE OF MY CLOSEST FRIENDS AND I KNOW IT HAS BEEN HARDER WITH MIXING IN BUSINESS. IVE ENJOYED THE CHALLENGES YOU AND ANDY AND I HAVE MET OVER THE YEARS IN GETTING THIS OFF THE GROUND.[3]

Weitbrecht agreed that it was difficult to end this chapter in their lives, but he knew it was best: **LIFE IS TOO SHORT TO BUTT OUR HEADS AGAINST A WALL.** Still, the business could not be closed down until Weitbrecht had a place to go. At his age, he was not interested in finding a position with another company. He also did not like the idea of merging with companies he believed had profited from his patent.

I AM JUST MAD THAT I HAVE NOT BEEN ABLE TO PRO- VIDE MORE EFFECTIVE PRODUCTS TO SELL . . . I ADMIRE ANDY AND I APPRECIATE HIS SERVICES VERY MUCH. HE HAS BEEN A GREAT HELP IN RUN- NING APCOM . . . HE IS ONE OF THE NICEST GUYS I HAVE EVER KNOWN AND YOU ARE ONE TOO.[4]

Saks was growing increasingly impatient with Weitbrecht's resistance to closing APCOM. An attorney had analyzed the firm's available options, including merging with other companies and an outright sale of APCOM. It was evident that APCOM no longer offered state-of-the-art products or had sufficient earnings to develop them. Saks called Marsters in late September: **I AM GIVING YOU A WARNING THAT I SHALL RESIGN FROM APCOM AND HAVE THE PREM- ISES VACATED BY FEBRUARY 1980 . . . WE ARE NOT GETTING ANYWHERE FOR THE PAST TWO YEARS AND IT HAS TO STOP.**[5]

At this point, Weitbrecht took a brief sabbatical from APCOM, renovated his house, and cleaned out some older equipment. "I am 60 in April," he wrote Chertok, "so I might as well get rid of the past so as to speak!"[6] It was time for changing the guard, but by March 1980, no decision had been made about APCOM.

APCOM was not alone in confronting the changes wrought by its own efforts. TDI, originally established to distribute surplus teletypewriters, had to adjust as electronic devices began to overtake the old TTYs and deaf people's demand for communication access spread. Telecommunications for the Deaf, Incorporated, became the organization's new name as it expanded interests to other forms of communication, including television captioning and computer technologies. Reflecting the new importance of political initiatives, its home office moved from Indianapolis to Washington, D.C. H. Latham Breunig, the founding president and an Indianapolis resident, resigned from his post as TDI's executive director. Breunig had nurtured a steady growth in the TTY network over a period of ten years, obtained tax-exempt status for TDI, edited the organization's newsletter, improved its public relations, and had given numerous presentations at legislative hearings and workshops, but now new struggles confronted the organization.

Incoming TDI executive director Barry Strassler believed that the principal challenge was to create awareness of the need for a nationwide telephone rate reduction for TTY users. In 1979, there were only two states with reduced TTY toll call rates. The first to act on this issue was Connecticut, inspiring TDI to believe that more states would follow.

Connecticut's leadership was sparked by the phone bills of a deaf woman, Barbara Babbini. Executive director of the Connecticut Commission on the Deaf and Hearing Impaired, Babbini was engaged to a man who lived in Colorado. For seven straight months, their combined long-distance charges totaled $400 a month, three to five times as much as two hearing customers would have to pay. Many other deaf people shared this concern about the cost of long-distance TTY calls. When a group of deaf people met with the Connecticut Public Utility

Commission, one of them lugged in a cumbersome Model 15 TTY and an acoustic coupler. He handed a page of text to a hearing member of the commission and asked him to read it aloud. He timed the reading with a stopwatch. He then gave the same text to a deaf printer, whose daily job included typing on a Linotype machine, and timed that individual as he typed on the TTY. Finally he asked another deaf person with average typing skills to type the message. The difference in the rates of oral reading and manual typing of the messages was clear to everyone.

This demonstration provided convincing evidence of the need for rate adjustments. Connecticut enacted a regulation allowing a 75 percent reduction in the telephone bills of deaf TTY users for intrastate calls. Although this action was a milestone in the story of the TTY, only 350 out of about 26,000 deaf people in Connecticut had TTYs at this time. Nevertheless, within a few months, several states followed Connecticut's example. In Missouri, Fred Banks's editorials in the local telecommunications advocacy group's newsletter were influential in leading to discount rates. The rate reductions varied greatly from state to state. In Kentucky and Tennessee, evening rates were applied for daytime TTY calls and night and weekend rates were used for evening calls. The New York Telephone Company applied an across-the-board reduction for local service as well as intrastate long-distance phone calls made by "certified deaf persons." Within two years, thirty-six states offered reduced rates to deaf consumers for intrastate calls, and AT&T filed a request with the FCC to reduce the interstate rates for deaf people using TTYs in their homes by 35 percent during the day and 60 percent during evening hours. TDI's GA-SK Newsletter enthusiastically announced, "Ma Bell Is Our Friend!"

It was ironic that at the very time APCOM was folding, AT&T was responding well to the needs of the deaf community. For decades, AT&T had been like a stubborn burro, refusing to move on the important expedition to telecommunications access. A breakthrough came during a forum called Telecommunications for Hearing Impaired Consumers, which brought together industry representatives and deaf people to discuss both technological and marketing developments.

The Organization for Use of the Telephone, whose efforts were spearheaded by David Saks, worked toward making sure that telephones were hearing aid compatible. The battle the APCOM partners had waged against an implacable telephone industry finally appeared to show signs of success. AT&T invited deaf consumers to help prepare a customer instruction booklet. The TDD Operator Services enabled deaf persons to call directly for assistance from AT&T. Eleven Bell operating companies were leasing telecommunications devices to customers.

TTY availability was moving forward quickly on many fronts. After nearly fifteen years of effort, in 1979 California moved on deaf people's demands for equipment access. State Senator William Green introduced a bill requiring that telecommunications devices be provided to deaf people at no cost and that service charges be high enough only to cover costs. The bill was drafted by the State Department of Rehabilitation, and in September 1979, Governor Edmund G. Brown, Jr., signed this landmark legislation requiring the Public Utilities Commission to implement a free TTY distribution program.[7] It was the first state law of its kind.

In Europe, by contrast, telecommunications for deaf people had advanced little in the eight years since Andrea Saks first traveled to London with two Phonetypes. Once her support from APCOM stopped in 1977, the movement became increasingly chaotic, and she found it difficult to assist the British deaf community in acquiring telephone access. By 1979, the United Kingdom had only 150 telecommunications devices in use by deaf people. One reason was the general economic climate in the United Kingdom. Phonetype couplers were not available at a price deaf people could afford. Furthermore, influential hearing people viewed the American Baudot TTY modem as a Trojan horse.

Saks saw years of her work lost when the Royal National Institute for the Deaf and Breakthrough Trust tried to modernize the small TTY network and create British portable telecommunications devices for deaf people. Influenced by recommendations from the Royal National Institute for the Deaf Technical Department and British Telecom, several electronic devices were manufactured and distributed over the next few years. Developed under a grant provided by the British Department of Health and Social Security, one device was touted as "the most advanced in the world," yet people who used it could only communicate with identical devices. Compatibility with Baudot TTYs was not endorsed, and the Phonetype couplers imported from America were made obsolete. Saks argued to no avail that existing telephone systems should not be thrown out. The decision to replace Baudot-compatible devices isolated deaf people in England from the TTY network in the United States and from the growing networks in Canada, Australia, and South Africa.[8]

The problem of incompatible telecommunication devices was spreading throughout Europe as other countries developed distinctly different technologies. In 1980, the West Germans were in the lead in Europe, with 900 portable terminals called the *Schreibtelefon*. This telecommunication device was compatible with those being proposed by Switzerland and Austria, but not with those in France or the United States. Meanwhile, the Dutch were exploring text telephones that decoded the input from push button telephones and required deaf people to use their voices. Sweden was offering "hard-wired" nonportable devices. Even twenty years later, no solution to the compatibility issues appeared imminent.[9]

In 1979, as APCOM was struggling and Weitbrecht's life seemed to be unraveling, he developed a new personal and professional relationship that would provide his remaining years with new meaning and would help to make TTYs available to many more deaf people. One of his deaf friends, the chemist Steve Brenner, had been working with him since 1974 on teleprinter and modem repairs. Brenner was particularly interested in bringing the price down for telecommunications devices, and he influenced several TTY designers in making more user-friendly, less expensive products. In Washington, D.C., in 1979, Brenner introduced Weitbrecht to Robert M. Engelke, a young hearing engineer from a Madison, Wisconsin, company called Automated Data Systems.

Weitbrecht and Engelke took to each other immediately. Engelke had developed a strong interest in assistive device technologies for people with disabilities. The previous summer at the National Association of the Deaf convention in Rochester, New York, he had demonstrated the VIP Communicator, which was a TTY the size of a hand-held calculator. The day they met in Washington, D.C., Weitbrecht and Engelke chatted for hours about networks and the future of TTY technology. They soon began to discuss collaboration on the design of new telecommunications devices.

Robert Engelke *(left)* gave Weitbrecht new directions in their collaborative research and development efforts after 1979. Courtesy of Robert Engelke, Ultratec.

A few months later, this new acquaintanceship began to reap direct rewards for Weitbrecht. Engelke established Ultratec, the first manu-facturer of TTYs to recognize Weitbrecht's patent and to pay him roy-alties. Engelke was more than a business associate for Weitbrecht, how-ever. He became a true friend, a beer-drinking buddy who challenged him to remain current with the new ASCII-based technologies. The computer was influencing the telecommunications movement at a rapidly increasing pace. Personal computers were being programmed to communicate with TTYs. Manufacturers of TTYs were planning ASCII terminal features to enable this relationship.

Ultratec's Superphone was among the first such advanced machines. It sold for about $395 at a time when other TTYs were averaging $600 to $800. Subsequently, several competing companies were forced to lower their prices. Weitbrecht was especially interested in Superphone because it included memory, ASCII code, and a microprocessor-based architecture. Learning more about the technology helped him under-stand the Dial-a-News ("Micro DAN") computer service he was work-ing on at this time. Superphone also had the capability of allowing another person who had no TTY to use a Touch-Tone phone to type short messages.

Pleased with this new machine, Weitbrecht began working with Engelke on the revolutionary Ultratec Minicom. Introduced in 1981 and offered for sale at only $159, the Minicom was the world's first low-cost, full keyboard TTY. Inexpensive and lightweight, it signifi-cantly broadened the opportunities for deaf and hearing people to communicate. Many smaller businesses were established to market the device. Affordability and availability were a reality at last!

Minicom's success demonstrated that, in the end, Weitbrecht, Saks, and Marsters not only wanted, but also needed, hearing people to benefit from TTY technology. This was the key to access and was be-hind Andrea Saks's efforts in Europe as well. From the beginning, she had argued that the hearing family members and friends of deaf people in England should also have access to TTYs. The more hear-ing people who used the new technology, the less expensive it would be for everyone.

The Ultratec Minicom was the first affordable TTY. Courtesy of Robert Engelke, Ultratec.

In 1982, Marsters, Weitbrecht, and Saks ended their partnership in APCOM. They had finally found a solution for Weitbrecht's retirement concerns when Engelke established a new company, Weitbrecht Communications, associated with Ultratec. It was formed for marketing and research. Weitbrecht was hired as chief engineer, and he also received royalties on each TTY Ultratec sold based on his patent. At the time that Marsters and Saks graciously bowed out of the TTY story, more than 100,000 telecommunication devices were in use by deaf people in the United States and many foreign countries.

From a business perspective, the APCOM partnership was not a financial success. When the newer electronic devices emerged in the early 1970s, APCOM did not capitalize on the opportunity to manufacture its own or to join forces with another manufacturer. Weitbrecht had already successfully tested the Phonetype modem for dual capacity to transmit and receive in Baudot and ASCII. He had the ability to build a more compact device, but by that time the market was expanding rapidly, and he felt defeated. The gentle, eccentric physicist turned the patent infringement issue into an obsession that consumed his time and energy over a period of years. It was left to others to turn the TTY technology into a lucrative business.

From a humanitarian perspective, Saks, Marsters, and Weitbrecht had significantly improved the quality of life for deaf and hard of hear-

ing people. For anyone who was involved with them, the financial benefits were minimal, but everyone benefited from the fact that telephone
access made nearly *anything* possible for deaf people. Their development of the Phonetype acoustic coupler paved the way for a technological revolution. The device and its successors opened doors to social, educational, and employment opportunities and facilitated a civil
rights movement toward anti-bias legislation. As *Business Week* magazine summarized shortly after APCOM closed, technology played a key
role in this movement and the "most pervasive piece of . . . electronic
equipment is the portable [TTY]."[10]

The telephone company played one last devastating game with
Weitbrecht. Although an evaluation committee recommended the
Ultratec Superphone for California's free distribution program, the telephone company offered the contract to Plantronics for the VuPhone.
The decision to go with VuPhone dismayed Weitbrecht. His work with
Engelke was to be his last chance to earn something for retirement
through royalties on sales. Ironically, the deaf community did not
widely embrace the VuPhone, and thousands were eventually scrapped
for five dollars each.

Engelke, too, was frustrated with the decision. He had just begun
his new company and the California distribution program would have
been a wonderful opportunity for growth. He was also angry that his
friend Weitbrecht had never received royalties from other manufacturers. The royalties earned from the California program would have
been a nice way to thank Weitbrecht for twenty years of dedicated work
toward access and equity for the deaf community.

In January of 1983, Weitbrecht decided to retire. Engelke offered
him 50 percent of his salary as a consultant in addition to the royalties
on his Baudot patent. The plan would allow Weitbrecht to live reasonably comfortably because it would bring him a steady income greater
than his current full-time salary. He prepared for a farewell lecture tour
to begin in September. He also began planning for a spring 1984 move
to a cabin in the Sierra Mountains and bought a new telescope. At
Weitbrecht Communications, he completed the software for his last

By the early 1980s, the TTY had become a common device in home, school, and community environments and it served as a bridge to the community. *(Top)* Kent Middle School in California began a TTY Pen Pal Club to encourage its students to exchange stories with other deaf children around the country. The children saved their money for long-distance TTY conversations with their friends. Courtesy of *World Around You,* Pre-College National Mission Programs, Gallaudet University. *(Bottom)* When WHEC-TV in Rochester, New York, sponsored its annual Muscular Dystrophy Association Telethon, Bonnie Meath-Lang, an English instructor at NTID, volunteered to take TTY calls from the deaf community. Courtesy of Bonnie Meath-Lang.

version of the Weitbrecht modem, which he and Engelke called "Phone-type 1000."

Throughout his life, Weitbrecht always had a dog by his side. They were his best friends. Blackie was his closest companion for thirteen years, often traveling in Weitbrecht's Chevy station wagon. When his dog Pan was hit by a car, the deaf physicist picked her up, cradled her in a blanket, and took her to the nearby veterinary clinic. The doctor could only reset a dislocated hip joint. Weitbrecht then took her to the Berkeley Veterinary Group, where, after listening to his pleas, a team of devoted staff finally pulled the dog through, at a cost of $400. "Now she is fine as ever . . . and quite lovable," he wrote Marsters.[11]

Weitbrecht's Irish Setter, Mickey, who Weitbrecht had found, was a "big fellow, very rambunctious and quite headstrong, just like me."[12] A neighbor helped Weitbrecht locate the owner, an unpleasant character who refused to reimburse the kennel fees Weitbrecht had paid while taking a short trip to Oregon on an eclipse expedition. Mickey, however, decided on his own where his future lay; he showed up several days after being returned to his original owner and eagerly leapt into Weitbrecht's lap. Weitbrecht often kept Mickey on a leash because the neighbors complained that the Setter picked fights with other dogs too often. Finding more time to be with his dogs and his telescope meant a great deal to Weitbrecht as he approached retirement.

On the evening of May 19, 1983, Robert H. Weitbrecht was walking Mickey in a crosswalk in Redwood City. A driver saw the dog and swerved in the dark, striking Weitbrecht. At the scene of the accident, Weitbrecht was having trouble breathing, and an unknown person expertly administered first aid, applying pressure to stop the flow of blood until paramedics arrived. Weitbrecht was taken to Sequoia Hospital in Redwood City. The police department attempted to find the man who slipped away in the rush of events, and thank him for "an unusual display of common sense and human compassion."[13]

The next day, no one had any idea where Weitbrecht was. At first, his friends were not concerned because he frequently disappeared with his dog in the midst of crises. When he took to the mountains, he often told no one where he was going. It was a coincidence that the

Robert H. Weitbrecht, 1920–1983. Courtesy of Sally A. Taylor.

day after the accident, APCOM's marketing manager Jeanne Poremba, still working with Weitbrecht in the new Weitbrecht Communications, had taken her daughter to the emergency room at Sequoia Hospital for a respiratory problem. Glancing at the admissions page, she noticed that an unidentified man had been admitted after an accident involving a dog. She ran up to the room to find Weitbrecht in a coma.

Poremba explained to the hospital staff that Weitbrecht was deaf. She then tried to communicate with him by tapping Morse code on the palm of his hand. She also tried flashing the hospital room lights. But none of these efforts succeeded. Weitbrecht died on Memorial Day, May 30, 1983, at the age of 63. His death shocked those who had come to appreciate his work in designing the acoustic telephone coupler. Following a beautiful ceremony, he was buried at Fair Haven Memorial Park in Santa Ana, California.

During the months that followed Weitbrecht's death, it was difficult for any deaf person who knew him to make a phone call without thinking of the physicist who had had such an impact on their lives. When the great deaf inventor Thomas Alva Edison died, President Hoover's momentary monument to his work was to turn off the White House lights for one minute on October 22, 1931. When Weitbrecht died, the light of another isolated genius went out for the deaf community he finally embraced.

13

LEGACY

Three of the four goals established by Marsters, Weitbrecht, and Saks had been met by the time Weitbrecht died in 1983. Mass marketing of compact telecommunications devices provided availability, portability, and affordability. The APCOM partners had also left the deaf community a legacy of self-advocacy, embodied by their example, the powerful national organization TDI had become, and the thousands of individual acts by deaf people across the country. All the strength of this legacy, and the experienced help of the attorneys working with the National Center for Law and the Deaf and the National Association of the Deaf, would be necessary to reach the final goal of accessibility through a national relay service.

The quality of telecommunications products for deaf people in the United States evolved rapidly in the 1980s. Strong demand was driven by a number of factors. Lower prices made the TTYs accessible to more individual deaf people. State distribution programs were important, as was the placement of TTYs in restaurants and other businesses. Government offices were major purchasers. The Federal Bureau of Investigation, for example, installed a telecommunications device to improve access to its services for deaf people, and the Federal Emergency Management Agency placed six TTYs in its regional offices. More and more state and federal government representatives installed TTYs in their offices for direct communication with deaf constituents. As competition grew intense, many of the two dozen smaller TTY manufacturers were driven out of business.

The problem of inadequate relay services for deaf people magnified quickly with several hundred thousand telecommunications devices in homes, offices, and businesses. The innovative Andrew Saks had established the first relay service in 1966. He had always found his own solutions to the relay problem over the years, usually costing him more time and money than should have been necessary. During a trip to Washington, D.C., for example, he requested hotel room service by using a portable TTY to call his private telephone relay service back in California and have them place his breakfast order by a long distance voice call. Such inconvenience and expense still characterized the daily experiences of deaf people. Now in retirement after years of dedicated voluntary service, Saks and Marsters both watched the telecommunications revolution from a distance. They looked forward to

The quality and availability of TTYs grew rapidly during the 1980s. Among the new models developed were *(clockwise)* the Krown Research Porta Printer Model MP-40, the Specialized Systems Incorporated System 100 Communicator, the American Communication Corporation TTY (from the collection of I. Lee Brody, courtesy of NY-NJ Phone-TTY, Inc., photographs by George Potanovic, Jr./Sun Studios), and the Ultratec Miniprint (courtesy of Robert Engelke, Ultratec).

the time when they could make calls to hearing persons without such undue efforts, but this time would not come easily.

An important figure in the push for relay service was Sheila Conlon Mentkowski. A deaf attorney with the National Center for Law and the Deaf at Gallaudet University, she had been tracking the distribution of telecommunications devices and identifying key issues associated with providing technical assistance to deaf persons. She had also charted reduced interstate and intrastate subscriber rates for telephone calls available in each state, and she was particularly interested in finding a way to have *all* states provide relay services similar to those then being offered on a voluntary basis in some states. During one committee meeting in the Maryland state legislature, Conlon Mentkowski, then seven months pregnant, testified that she considered it humiliating to have to rely on volunteers to call her family to let them know of something as intimate as the birth of a child. Like many other states, Maryland viewed relay services as a form of charity rather than a public utility.

Conlon Mentkowski made her point at the federal level as well. At the end of 1986, the FCC was finally turning its attention to the telecommunications barriers facing people with disabilities, and Conlon Mentkowski was scheduled to be a witness at a December 5, 1986, hearing. For this meeting, she had to contact the individual in charge of planning to inform him that she needed a sign language interpreter, but it was not easy for her to telephone him. There was no TTY in his federal building. After finally reaching him and requesting an interpreter, he told her to bring her own. He later relented but then arranged for only one interpreter to work for the four hours of scheduled testimony. With prominent representatives of the telephone companies looking on, Conlon Mentkowski recounted the difficulties she experienced with both the telephone and getting an interpreter to attend just this one hearing. She argued the critical point that one of the most serious challenges in establishing a relay service was convincing hearing people that it was a human service—not specifically funded for deaf persons and those with speech disabilities—but for *everyone*.

Maddeningly, the committee at first just did not get her point that communication access was essential. Midway through the meeting, the lone interpreter needed a break. As the interpreter began the break, the FCC representative began to cover other matters. Karen Peltz Strauss, a hearing staff attorney for the National Center for Law and the Deaf, and several others shouted out that no business should be conducted while the interpreter was not there. They were denying the deaf attorney access to the inquiry. In spite of this incident, the meeting was nevertheless productive and eventually led to the FCC's Notice of Inquiry on Access to Telecommunications Equipment. Among other things, this proceeding sought public comment on the development of an interstate relay system—designed to handle relay calls among the states. Six months later, the National Association of Regulatory Utility Commissioners submitted a petition to the FCC requesting an investigation of the implementation of a nationwide telephone relay system. Meanwhile, states continued to work on solutions, and new problems soon became apparent.

California took a major step toward fulfilling the remaining goal established by the APCOM partners on New Year's Day, 1987, when a statewide relay service began in Woodland Hills, California. A new California Law, Bill SB 244, required the Public Utilities Commission to design and implement, over a three-year period, a program whereby telephone companies would provide a dual-party relay system using a third-party operator to connect TTY users with voice calls. The resulting AT&T twenty-four-hour, seven-day-a-week California Relay Service provided more than 100 staff members trained to relay telephone messages. Within a few years, the California Relay Service grew to include more than 250 "communications assistants" and handled an average of 230,000 phone calls each month.

As many other states established professional relay services, the need for access overwhelmed the personnel. When Virginia Relay Service in Charlottesville opened in July 1987, it had only 198 customers. Within a year, this number soared to 2,400, and the service had to be shut down because of the high cost of operation. Many other systems with professional staff were also struggling. In 1987, New York State had not yet set up a statewide relay system. The burden on local services was overwhelming. Hi-Line Relay Service in Rochester, for example, was handling 163,497 calls with only four operator positions. There was a blocking rate (TTY callers getting busy lines) of 30 percent.

In California, costs quickly overwhelmed expectations. When the California relay program began in January 1987, a three-cent surcharge was included in everyone's telephone bill to cover the program's expenses. By October, insufficient funds led to an emergency increase to ten cents, effective January 1, 1988. Six months later, the surcharge was again raised, and over the next decade the problems continued. All the while, deaf people lived in fear of having the relay service shut down.

Poor training of personnel responsible for providing services provided another source of frustration and worry. In October 1986 in San Diego, California, for example, a deaf woman died when a local 911 emergency number was not properly answered by TTY. It took several hours for emergency personnel to respond to her husband's frantic calls. A television channel followed up with a filmed attempt to place a local 911 call to the Sacramento County Sheriff's Department. It required two tries to complete the call. The person who answered was not familiar with the procedure for transferring the call from the switchboard to the telecommunications device.

Due in large to part to the interest of Senator Tom Harkin of Iowa, whose brother is deaf, opportunities to bring these issues to the attention of Congress were increasing. In 1987, the staff of the National Center for Law and the Deaf reviewed strategies to exploit the new political environment. One issue discussed was how to expand state programs that distributed free telecommunications equipment to deaf people. Another concern was the need for reduced rates for long-distance relay service calls. The concept subsequently was introduced to the Senate Labor and Human Resources Committee. Sheila Conlon Mentkowski believed that the time was right for a federal law requiring all states to provide for unrestricted calls by deaf and hard of hearing consumers twenty-four hours a day, seven days a week. She and her colleague,

Karen Peltz Strauss, then followed up with numerous meetings with people with disabilities to begin designing federal legislation.

Over the next several months, Peltz Strauss and Conlon Mentkowski worked closely with Alfred Sonnenstrahl, a deaf engineer who was TDI's fourth executive director; Paul Taylor, a deaf professor on leave from the National Technical Institute for the Deaf; and representatives from consumer organizations to draft preliminary language for the federal bill that would mandate such services. The goal was to create a comprehensive relay system with permanent funding that would provide equitable access to telephone services for deaf people. In the past, the limited funding appropriated by many states made equitable access subject to re-evaluation with each state budget.

By January of 1988, Peltz Strauss and Conlon Mentkowski had completed a draft of a federal bill requiring nationwide relay services and sent it to Robert Silverstein in the U.S. Senate Committee on Labor and Human Resources. They explained that AT&T had not provided such a service because it was waiting for both guidance and authority from the FCC. "Indeed, AT&T has been very interested in establishing a nationwide program for some time."[1] At this time, only eight states had TTY relay systems in operation and about thirteen were in the process of studying the development of such services.

Peltz Strauss and Conlon Mentkowski also sent the draft to Taylor. He wrote to the two attorneys with excitement: "What a pleasant surprise to hear from you and to see the bill you have been working on. . . . Not even in my wildest dreams did I imagine that progress on the telephone relay service would escalate to the federal level [so quickly]." As a college professor, Taylor also saw the potential impact on new and current employment opportunities for graduates of NTID, Gallaudet University, and other postsecondary programs. "The sooner the relay services become unconditionally accessible, the much greater the number of jobs will become available for the graduates now that the telephone barriers have come down."[2]

Deaf rights generally, including telecommunications access, captured world attention in early March 1988 with the "Deaf President Now" protest at Gallaudet University in Washington, D.C. Students, staff, faculty, and alumni closed down the university for a week as they demanded the resignation of Elisabeth Zinser. A hearing administrator who did not know American Sign Language, Zinser had been selected the university's president over two deaf finalists. The university's board of trustees, confronted with this dramatic message about deaf self-empowerment, eventually relented, accepted Zinser's resignation, and appointed I. King Jordan, one of the deaf candidates, as Gallaudet's first deaf president. As Senator Tom Harkin summarized a few months later, "No one will give Americans with disabilities their rights—you will have to fight for them. And you must fight, because your civil rights have not yet been secured."[3]

It was fitting that the TTY played an important role in the urgent communication that took place during this protest. Sonnenstrahl had donated twenty-five copies of TDI's Blue Book TTY directories to help the protesters establish and maintain lines of communication nationally. Deaf President Now leaders and supporters used TTYs to raise funds, contact the media, organize supporting demonstrations elsewhere, and keep the national and international deaf community apprised of the revolt's progress. The TTY had become both a means and an end for deaf activism.

In the second half of 1988, as informal discussions continued to take place about a nationwide relay service, consumer groups began working on legislation for a federal relay system as well. The latter legislation focused on ensuring that the federal government telecommunications system would be fully accessible to deaf and hard of hearing people. This would require relay services to, from, and within federal government offices and the distribution of telecommunications devices to offices of members of Congress. Working toward this goal in June 1988, Sonnenstrahl and Taylor represented TDI at a Senate hearing chaired by Senator Daniel Inouye. Sonnenstrahl also took the opportunity to push for an international TTY logo to be part of the law. The bill passed unanimously, and a companion House version was introduced by Congressman Steven Gunderson of Wisconsin.

The National Association of the Deaf (NAD) contributed to this political effort, and legislation continued moving in Washington, D.C.

The international TTY logo was created as part of the Telecommunications Accessibility Enhancement Act of 1988. The law established an expanded federal relay service. Courtesy of Telecommunications for the Deaf, Inc.

At NAD's convention in Charleston, South Carolina, Conlon Mentkowski stood before a large audience and described actions by the federal government to improve telecommunications access for people with disabilities. More than a thousand signatures in support of the federal relay legislation were then collected. While Title IV of what became the Americans with Disabilities Act was being developed in collaboration with Senator Harkin's staff, Senator John McCain of Arizona sent a separate bill to the Senate floor promoting a national, federal relay service. The bill, co-sponsored by Senators John Danforth and Robert Packwood, was approved by the Senate and sent on its way to the House.

The battle for the federal relay service was not over, however. The deaf community learned that, due to a backlog, the bill would likely die in a House committee. In October 1988, NAD and TDI representatives pushed for legislative action. Students from Gallaudet University and the National Technical Institute for the Deaf, representatives from the Deaf Section of the Federal Employees' Organization, Senate supporters, House supporters, and the Washington, D.C., deaf community converged on the Capitol to protest. In the wake of the Gallaudet University protest a few months earlier, the rally proved effective. The House Committee chairpersons and the House Speaker met with seven groups from the rally, and HR 4992 appeared on the House floor a few days before adjournment. It was passed. Congressman Steve Gunderson summarized this group effort by quoting Gallaudet University President I. King Jordan, calling it a "helpful beginning to finally achieving equal telephone access for all Americans."[4]

In October, President Ronald Reagan signed the Telecommunications Accessibility Enhancement Act of 1988, establishing an expanded federal relay service for calls to, from, and within the federal government. Its purpose was to improve telecommunications access for TTY users. Earlier that year, Congress had also passed the Hearing Aid Compatibility Act, which required that all landline telephones (with the exception of secure telephones) made in or imported into the United States must be hearing aid compatible.

After passage of the Telecommunications Accessibility Enhancement Act of 1988, energies were turned toward the nationwide relay services. During this period, states also continued their individual efforts to provide relay services to their residents. The establishment of each state relay service required an incredible amount of planning. In New York State, forty-one telephone companies, AT&T, and deaf consumers were involved in establishing the New York Relay Center. With funding from the telephone companies, the center opened in Clifton Park near Albany in January 1989, and more than 44,000 calls were made that first month. Within one year, the monthly total had increased to more than 101,000 calls. After another year, the volume of TTY calls had increased to more than 1.3 million per month. In other states, such as Michigan, a single local exchange carrier agreed to provide relay services throughout the state. Despite these efforts, there were not more than sixteen states with operational state relay systems, and most were plagued with funding shortages, blocking rates, and restrictions on the length of time a deaf person could talk on the telephone, the number of calls that could be relayed, and the type of calls that could be made. Arkansas, for example, limited users of its relay service to fifteen minutes per call and did not permit personal calls to be made. Emergency phone calls in Massachusetts could be made only between 7 A.M. and 11 P.M. Some states did not permit interstate calls.[5]

The problems confronting state relay programs helped to highlight even further the need for a federal relay mandate. As plans for the Americans with Disabilities Act (ADA) developed, the telecommunications section, called Title IV, evolved from the 1988 draft of the bill that encouraged "local telephone companies to provide dual party relay systems for the deaf and hearing impaired" to mandated legal requirements setting the scope and responsibilities of relay service providers.[6] But even as it became clear that these relay provisions would become law, many consumers remained concerned about TTY access in emergency situations. The planners knew that the extra time needed to handle relay calls could be life threatening during an emergency. One solution, sought by NAD and TDI, was to include language in the ADA requiring TTYs to be placed in all 911 centers. But with progress on the ADA occurring rapidly, Congress appeared reluctant to add a new section to the legislation. Peltz Strauss and Robert Silverstein came up with a solution. They looked at Title II and saw that it prohibited discrimination by city services. It was clear to them that 911 was a city service, so it would not be permitted to discriminate against TTY users. Rather than draft a new legislative section, Peltz Strauss drafted report language, to be included in the legislative history of the ADA, which stated that *direct* TTY access, not relay services, was needed in the event of emergencies.[7]

With the 911 matter resolved, all was in place to push the relay section through Congress. Senator McCain and Congressman Edward J. Markey took on this challenge. It was through their efforts, the advocacy of TDI, the NAD, and the legal assistance of Peltz Strauss that the provisions for relay services in Title IV of the Americans with Disabilities Act came about.

Sadly, on May 6, 1989, at the height of the movement for access to national telephone relay services, Andrew Saks died of a heart attack. His dedicated support to the telecommunications network for deaf people in general and for the well-being of APCOM, in particular, helped the historic TTY movement to grow. Saks was a behind-the-scenes figure who never took the spotlight. From spray painting the original modems in his own garage to traveling through Canada and other countries with his wife, Jean, to demonstrate the TTY, he was deeply involved with building the grassroots telecommunications network. He established the first telephone relay service, which enabled deaf people with modems to have telephone communication with anyone, with or without TTYs. Had this quiet member of the inventive trio from APCOM lived one more year, he would have proudly witnessed the signing of the Americans with Disabilities Act, the culmination of the telephone relay service he had conceived twenty-five years earlier.

Andrew Saks, 1917–1989. Courtesy of Jean M. Saks.

When President George Bush signed the Americans with Disabilities Act in 1990, he was joined by *(left to right)* Evan Kemp of the Equal Employment Opportunities Commission, Sandra Perrino of the National Council on Disability, and Justin Dart of the President's Committee on Employment of People with Disabilities. Courtesy of Stephen A. Brenner.

The House of Representatives voted overwhelmingly (403 to 20) for ADA legislation to expand legal protections and rights for people with disabilities on May 22, 1990. The Senate had passed a similar version earlier. President George Bush signed the ADA into law on July 26, 1990. Sonnenstrahl represented TDI at the White House ceremony and personally thanked President Bush and the ADA's cosponsors, Senator Harkin and Congressman Steny H. Hoyer. In TDI's newsletter, Sonnenstrahl reflected on the past three decades since Weitbrecht, Marsters, and Saks implemented the TTY network with the Phonetype acoustic telephone coupler. In 1964, people witnessed the first TTY phone call between two deaf persons. In 1990, they witnessed the signing of the ADA. These were, in his opinion, two of the greatest years in the history of telecommunications access for deaf and hard of hearing Americans.[8] The "universal service" mandate had finally become a reality for them.

Following passage of the ADA, there was much work to be done to implement Title IV. Peltz Strauss served as legal counsel for more than seventy local and national consumer groups in drafting comments to the FCC on how best to implement the new relay service mandates.

Marvin Arend, the inventor of the pay telephone kiosk, watched as TDI Executive Director Alfred Sonnenstrahl demonstrated the kiosk during the TDI convention in Anchorage, Alaska, in 1993. Courtesy of Telecommunications for the Deaf, Inc.

At the Technology Assessment Program at Gallaudet University, Judith E. Harkins and her staff assisted in setting telecommunications standards and telecommunications training. AT&T sought ways to respond to the needs of deaf customers, marketing a new family of telecommunications devices that incorporated printers, answering machines, and personal directories into portable units. TDI established an Emergency 911 Access Team that offered unique expertise and training materials to emergency service providers. Under Sonnenstrahl's leadership, TDI increased its national directory listings from 9,000 to 55,000. The organization also developed the Operator Assisted Service program for users of telecommunications devices and helped Northwestern Bell and Chesapeake and Potomac Telephone companies meet the needs of deaf customers.[9]

Telephone relay services, the fourth and final goal established by Robert Weitbrecht, James C. Marsters, and Andrew Saks, became a reality because of the contributions of many deaf and hearing persons through the years. By July 1993, all fifty states had established relay systems that operated twenty-four hours a day, seven days a week, with complete confidentiality, no calls rejected, and no restrictions on length of calls or their content.

In an important celebration of these efforts in July 1993, President Bill Clinton sat down to place a demonstration call to Frank Harkin (Senator Tom Harkin's brother) in Cummings, Iowa, to inaugurate the Federal Information Relay Service. The president could not get through; the line was busy. He tried again; and then a third time.

Senator Harkin's spirits were momentarily dampened. He had a long history of involvement in breaking down technological and attitudinal barriers to enhance the lives of deaf and hard of hearing people. His brother, Frank, had become deaf from meningitis in childhood and later helped assemble the very Navy jets Tom flew in the 1960s. When Harkin was sworn in as a United States Senator in January 1985, he convinced the Sergeant-at-Arms to provide a sign language interpreter for his brother in the Senate Gallery. It was the first time this had ever been done. Harkin struggled for six years to have the Senate proceedings closed captioned with real-time captioning;[10] in July 1990, when the Senate passed the conference report accompanying the Americans with Disabilities Act, he spoke to his brother and millions of other deaf and hard of hearing persons in sign language without voice, thanking Frank for teaching him at an early age that people with disabilities could do anything they desired.

Now, at the inauguration of the relay system in July 1993, disability leaders watched President Clinton dialing as fifteen cameras and the print press captured the moment. An anxious Senator Harkin had to think quickly. He called a neighbor in Iowa to run next door to tell his brother to get off the phone because the president of the United States was trying to call him. Frank and the president then had a delightful conversation through a telephone relay operator. The modern era of the telephone for deaf people had finally arrived.

APPENDIX

A Concise History of the TTY

1837   Samuel F. B. Morse demonstrates the telegraph, the first electrically operated machine for distance communication.

1874   Thomas Alva Edison patents the duplex telegraph, which allows two messages to be transmitted simultaneously over the same wire. Emile Baudot develops a five-level telegraphic coding system.

1876   Alexander Graham Bell demonstrates his voice telephone at the Centennial Exhibition in Philadelphia.

1893   Elisha Gray's Telautograph, an electric writing machine for use with the telephone, is demonstrated at the World's Fair in Chicago.

1912   William E. Shaw demonstrates the "Talkless Telephone."

1920s   Bell Telephone Laboratories established.

       Bell System creates the "Deaf Set" for hard of hearing persons.

       Bell System demonstrates transmission of pictures over telephone lines.

1934   Congress passes the Communications Act, which includes a provision requiring the recently established Federal Communications Commission to ensure "universal services . . . so far as possible to all the people of the United States."

1957   Bell Laboratories demonstrates a TV-Telephone.

1964   Visual Speech Indicators are developed. These hand-held devices are equipped with a moving needle that indicates whether someone is speaking on the other end of the telephone.

       The first long-distance call by deaf persons using electric writing machines occurs between the Vocational Administration Office in Washington, D.C., and the San Fernando Valley State College Leadership Training Program in California.

James C. Marsters recommends TTY communication over regular telephone lines.

Robert H. Weitbrecht develops an acoustic telephone coupler for use with teletypewriters by deaf people.

National Association of the Deaf convention in New York City includes exhibits of telephone devices.

First public demonstration of a TTY call between deaf persons takes place in a hotel at the Alexander Graham Bell Association for the Deaf convention in Salt Lake City, Utah.

AT&T demonstrates the Picturephone at the World's Fair in New York City.

Victor-Comptometer Corporation Electrowriter is used at the World Games for the Deaf in Washington, D.C.

Robert H. Weitbrecht in Redwood City, California, places the first long-distance TTY call to James C. Marsters in Pasadena, California.

1965 Robert H. Weitbrecht Company partnership formed; first "Gray Lot" modems are built by Weitbrecht, James C. Marsters, and Andrew Saks.

Carterfone case stalls distribution of TTYs.

First transcontinental TTY call takes place between Robert H. Weitbrecht in New York and James C. Marsters in California.

Andrew Saks suggests relay telephone service concept.

Robert H. Weitbrecht experiments with the "voice carry over" method based on suggestions from Andrew Saks and James C. Marsters.

1966 James C. Marsters lectures on the TTY technology breakthrough to deaf communities in Europe.

Andrew Saks establishes the first telephone relay service in Redwood City, California.

James C. Marsters establishes the second telephone relay service in Pasadena.

Robert H. Weitbrecht files a patent for the "Frequency-Shift Teletypewriter."

Weitbrecht makes a demonstration TTY call to Marsters from the Vocational Rehabilitation Administration to gain government support for the technology.

Eighteen TTYs are in use by the end of the year.

1967   Stanford Research Institute holds a planning meeting to discuss telecommunications needs of deaf persons.

Applied Communications Corporation (APCOM) is established to manufacture the Phonetype modem.

Paul L. Taylor establishes the first local telecommunications group, the Telephone/Teletype Communicators of St. Louis.

1968   The eight-level American Standard Code for Information Interchange (ASCII) is defined by the American National Standards Institute as the federal standard for computer data transmission.

Carterfone case is settled by the FCC; the ruling permits consumers to connect all manufacturers' equipment to telephone company lines.

Telephone/Teletype Communicators of St. Louis establishes the third local telephone relay service.

AT&T reaches an agreement with the Alexander Graham Bell Association for the Deaf to distribute TTYs.

Teletypewriters for the Deaf Distribution Committee (TDDC) is established by the National Association of the Deaf and the Alexander Graham Bell Association of the Deaf; TDDC is renamed Teletypewriters for the Deaf, Inc. (TDI) in June; H. Latham Breunig becomes TDI's first executive director.

Stromberg-Carlson Vistaphones are field-tested at the National Technical Institute for the Deaf at Rochester Institute of Technology.

The first TTY weather service and TTY news service are established in St. Louis.

174 TTYs are in use by end of the year.

1969   National Technical Institute for the Deaf begins a research study to design portable TTYs for use by deaf consumers.

First international TTY call takes place on January 4, between deaf people—Robie Scholefield in Vancouver, British Columbia, and Vicki Hurwitz in St. Louis, Missouri.

ESSCO Communications and Ivy Electronics introduce competing modems.

First transatlantic call is made between two deaf persons using video telephone technology (AT&T's Picturephone).

600 TTYs are in use by the end of the year.

1970   Weitbrecht's patent for the modem is approved by the U.S. Patent Office.

APCOM begins marketing the Automatic Control Unit answering device for unattended TTYs.

First intercontinental (transpacific) TTY call is placed between Minneaplis, Minnesota, and Manila, Philippines.

ESSCO ATC-2 becomes the first modem to compete with the Phonetype.

U.S. Government Printing Office in Washington, D.C., installs a TTY for deaf employees.

FCC permits connection of devices not provided by the telephone company to the telephone network.

900 TTYs are in use by the end of the year.

1971   Internal Revenue Services rules that the costs of specialized equipment (including acoustic telephone modems) is deductible as a medical expense.

Hotline for the Deaf is established in Maryland.

First National Conference of Agents of Teletypewriters for the Deaf, Inc., is held in Washington, D.C.

TTYs are installed in police departments in Dallas and Los Angeles, the first efforts to provide emergency assistance to deaf people.

New York-New Jersey Phone-TTY introduces the first accessible Dial-A-News Service.

1,500 TTYs are in use by the end of the year.

1972   Microminiaturization of electronic circuits leads to lighter and quieter devices manufactured by HAL Communications Corporation and MAGSAT.

St. Louis begins transmitting news stories from UPI wire feeds.

A TTY is installed at a TV station (KRON-TV, San Francisco) for call-ins by deaf viewers.

Andrea J. Saks brings the Phonetype modem to London, England.

2,500 TTYs are in use by the end of the year.

1973   New York Telephone, Indiana Bell, and New Jersey Bell are the only telephone companies to waive unlisted number charges for TTY users.

The Rehabilitation Act is signed by President Richard Nixon.

David Saks establishes Organization for Use of the Telephone (OUT) to address the needs of hearing aid users.

After four years of testing, NTID's Vistaphone (videotelephone) is discontinued due to bandwidth problems.

3,000 TTYs are in use by the end of the year.

1974   First International Convention of Teletypewriters for the Deaf, Inc., is held in Chicago.

4,800 TTYs are in use by the end of the year.

1975   First authorized transatlantic TTY call is made between England and the United States.

A three-way TTY call is made during the World Federation of the Deaf Congress by callers in Washington, D.C., San Francisco, and Sweden.

The first statewide, toll-free relay service is established in North Dakota.

10,000 TTYs are in use by the end of the year.

1976   FCC Commissioner Richard E. Wiley installs a TTY in the Consumer Assistance Office.

20,000 TTYs are in use by the end of the year.

1977   600 families with TTYs in the Philadelphia area begin receiving news through radio receivers.

National Center for Law and the Deaf comments on TTYs in public facilities are filed with the U.S. General Services Administration.

35,000 TTYs are in use by the end of the year.

1978 Pacific Bell establishes statewide centers in California to provide technical assistance to people with disabilities.

1979 Governor Edmund G. Brown of California signs landmark legislation for distribution of free TTYs.

Connecticut becomes the first state to reduce long-distance rates for deaf people.

A TTY is installed in the Old Executive Office Building next to the White House for President Jimmy Carter's "Comments Office."

Barry Strassler is appointed the second executive director of TDI.

TDI changes its name to Telecommunications for the Deaf, Inc.

1980 Electronic messaging (e-mail) experiments are conducted with DEAF-NET in Washington, D.C., and San Francisco, and with HERMES in Boston.

AT&T establishes toll-free TTY operator services.

Twelve states allow reduced rates for intrastate long-distance TTY calls.

California begins free TTY distribution program for deaf residents.

1981 AT&T files a request with the FCC to reduce rates for interstate TTY calls.

Electronic Industries Association begins efforts to develop standards for TTY manufacturers.

More than 30 states provide reduced rates for long-distance TTY calls.

1982 APCOM closes.

Congress passes the Telecommunications Act of 1982; the law expands telephone access for people with disabilities, based on the universal service obligation.

180,000 TTYs are in use by the end of the year.

1983   AT&T petitions state commissions to remove tariffs on special telephone equipment for deaf and hard of hearing persons.

AT&T establishes the Nationwide Telecommunications Devices for the Deaf Center in Reston, Virginia, to meet the special long-distance telecommunications needs of deaf and hard of hearing customers as well as people with speech-related disabilities.

Canada gives $600 vouchers to deaf people to purchase TTYs.

1984   Federal government investigates placement of TTYs in public transportation facilities.

Thomas M. Mentkowski is appointed the third executive director of TDI.

AT&T Special Needs Center is established in New Jersey.

1985   Low-cost TTYs and dual TTY/ASCII modems become more available.

More states provide TTY distribution programs.

1987   AT&T offers the first public telephone relay service in Woodland Hills, California, to comply with a California state law mandating access to all telephones within the state; 80,000 calls are made in the first month.

Alfred Sonnenstrahl is appointed fourth executive director of TDI.

Ultratec produces a dual TTY/ASCII Intel modem for DOS and MacIntosh computers.

1988   President Ronald Reagan signs the Telecommunications Accessibility Enhancement Act and the Dual Party Relay Service Act.

Ultratec pay-phone TTYs are installed in airports, schools, and other public locations.

1989   President George Bush makes the first call on the expanded federal relay service.

Judge Harold Greene waives long-distance restrictions for the "Baby Bells" for relay services.

1990   President George Bush signs the Americans with Disabilities Act of 1990.

Advances in fiber optic technology improve research developments in video telephones.

New York-New Jersey Phone-TTY develops software allowing automatic billing for relay services.

1991   Federal relay standards are defined by Title IV of the ADA.

1993   TDI begins developing TTY equipment standards.

Title IV of the ADA takes effect.

US Sprint is awarded the contract to operate the Federal Information Relay Service.

1994   National Association for State Relay Administrators is established.

MCI offers the first calling card for TTY users.

1995   Sprint experiments with video relay interpreting in Texas.

1996   President Bill Clinton signs the Telecommunications Reform Act of 1996, directing the FCC and a joint board of state and federal communications regulators to reexamine the concept of universal service.

1997   Claude L. Stout is appointed the fifth executive director of TDI.

1998   TDI celebrates its 30th anniversary.

2000   TDI formally shortens its name from "Telecommunications for the Deaf, Incorporated," to "TDI."

NOTES

Introduction

1. R. H. Crutchett, "Call Me Up Sometime," *Volta Review* 37 (1935): 101.
2. For additional discussions of TTY terminology, see *GA-SK Newsletter*, 10, no. 2 (1980); 12, no. 2 (1982); 23, no. 2 (1992); 23, no. 3 (1992); 23, no. 4 (1992); 24, no. 2 (1993); and *Deaf American*, 32 (July–August 1980): 30.
3. Sylvia Schechter, letter to author, January 16, 1996.
4. Allan F. Bubeck, "How to Use the Telephone in Emergencies," *The Silent Worker* 13 (July 1961): 23.
5. Robert G. Sanderson, letter to author, February 1995.
6. Samuel C. Florman, *Blaming Technology*, 93.

1. A Chance Encounter

1. Teleprinters were available as early as the 1840s. The "printing telegraph machine" recorded messages in Roman letters. In 1849, the teleprinter was used to send messages between New York and Philadelphia. Various forms of the telegraph also came out, many of them offering potential convenient access for deaf people. The "autographic telegraph" in 1848, for example, allowed a person to write a telegram on a piece of paper. This message could be transferred to a metal plate for transmission over telegraph lines. Some entrepreneurs envisioned this technology in every American home and business.
2. Robert H. Weitbrecht, "The RTTY Story at W6NRM/W9TCJ," *RTTY* 4, no. 7 (July 1956): 5.
3. Robert H. Weitbrecht, "W6NRM/RTTY at the San Mateo Hamfest," *RTTY* 7, no. 8 (August 1959): 8.
4. Ibid., 9.
5. Robert H. Weitbrecht, "Tape Off the Floor," *RTTY* 7, no. 2 (February 1959): 11.
6. During his career, Weitbrecht combined his photography and astronomy interests, producing one of the first designs for automatic camera

systems for Dearborn Observatory, and a high-precision astrometric camera system for Lick Observatory. He also conducted studies of high-definition photography for the ECHO satellite system.

7. James C. Marsters, letter to Robert H. Weitbrecht, April 3, 1964.

8. Marsters, letter to Weitbrecht, April 26, 1964.

9. Weitbrecht, letter to Marsters, April 29, 1964.

10. Marsters, letter to Weitbrecht, April 30, 1964.

11. Marsters, letter to Weitbrecht, April 3, 1964.

2. Up the Mountainside

1. Fred DeLand, "The Telephone, the Radiophone, the Graphophone, the Music Record, and Modern Lip-Reading," 251. Bell's interest in developing the telephone diaphragm was sparked by his wife Mabel's comment that she could feel sounds through her muff while walking in the winter weather.

2. "The Telautograph," *Our Deaf and Dumb* 1 (March 1894): 111–12. Deaf people were able to send telegraph messages through Western Union service just like hearing persons, although the telegraph was primarily a means for business calls and was used much less for personal exchanges. One could not easily telegraph a doctor or a neighbor.

3. *Silent Worker* 12 (April 1900): 119. The failure of deaf people to protest the decision to discard use of the Telautograph at the turn of the century is due to a number of factors. First, businesses monopolized the telephone during its early years, and the Bell Company was very responsive to their needs. The company was much less influenced by the priorities of "social telegraphy." By the turn of the century, there were more than two million telephones in the United States, but only a small fraction were found in residential environments. Most telephones were being used by large and small businesses.

Another reason for the lack of reaction was economic. As early as 1882, a residential telephone in New York City cost $150 a year. Because most deaf persons were in low-income occupations at this time, the cost was far beyond what they could afford. Social factors also played a large role in deaf people's acquiescence. These factors included the emotional debate over communication methods; the dismissal of deaf teachers by school administrators who expected all instructors to teach speech; the Eugenics movement, whose proponents sought to discourage marriages between deaf people in order to eliminate hereditary deafness; the refusal of many states to license deaf drivers; and the difficulty of obtaining life insurance. It is no wonder, then, that the pursuit of the telephone, or any technology intended to make life more comfortable, was second to the need to just survive in a paternalistic society.

4. "Telephone for Deaf-Mutes," *California News* 28 (October 1912): 7.

Nothing more was heard of this invention. Several major contributors to the science of telecommunications in the nineteenth and twentieth centuries experienced deafness through much of their careers as scientists or inventors, yet they did not pursue a visual form of telephony at any length. These included the American inventor Thomas Edison, Germany's Ernst Werner von Siemens, and the British electrical scientists Oliver Heaviside and John Ambrose Fleming. See Harry G. Lang and Bonnie Meath-Lang, *Deaf Persons in the Arts and Sciences*, for biographies. See also Harry G. Lang, *Silence of the Spheres: The Deaf Experience in the History of Science*.

5. Henning C. F. Irgens. "Telephone Communication Experiments," in *Report of the 42nd Meeting of the Convention of American Instructors of the Deaf* (1965): 55–56.

6. The President's Committee on Employment of the Handicapped, *New Invention Brings New Hope for the Deaf* (Washington, D.C.: U.S. Government Printing Office, 1964), 14. 7. The Speech Indicator is mentioned in the *Deaf American* 17 (February 1965): 11, and 18 (July–August 1965): 21. Weitbrecht corresponded with the engineer Gale H. Smith at AT&T (July 13, 1965) about the relative merits of the yes-no indicators. They also discussed the Regency Electronics Code Translator designed for the armed forces, known as Codamite. The system included a keyboard, and Weitbrecht wondered if it could be used to send and receive signals with some form of electronic alphanumeric readout. Codamite, however, was too complex electronically. Its circuit included 350 diodes, 75 transistor circuits, and 17 small incandescent lamps.

8. Robert H. Weitbrecht, "Recent Experiences in the Field of Communication Devices for the Deaf," 4.

9. The Bell system has a long history of assisting hard of hearing people through research; one such development was the "Audiphone," an early form of the vacuum-tube hearing aid, that had a microphone transmitter attached to clothing and a receiver placed on a table. As Bell Labs scientists investigated the properties of speech and hearing, however, the emphasis on increasing medical knowledge became a predominant goal.

10. Many deaf people were denied positions or advancement because they were unable to use the telephone. See Harry Best, *Deafness and the Deaf in the United States*, 227, and Alan B. Crammatte, *Deaf Persons in Professional Employment*.

3. Something Old, Something New

1. The five-level Baudot code was capable of transmitting 64 characters. ASCII allows the transmission of 128 or more characters and at a much higher rate.

2. Margaret Stovall, "Teletype Breakthrough for the Deaf." Although a November date was reported in this article, many earlier attempts had been made.

3. Weitbrecht's original patent application for the "Frequency-Shift Teletypewriter" was filed August 22, 1966. The invention was an improved system that could be used over the telephone line, overseas cable, satellite, or radio. The key features were the reduction of signal noise through an amplifier and filter and the use of an acoustic coupler in the form of a cradle. The patent was formally approved on April 21, 1970. An application for reissue was made on April 5, 1977, and the patent was reissued on March 6, 1978.

In addition to tone frequencies of 1400 Hz (Mark) and 1800 Hz (Space), Weitbrecht's first acoustic coupler was designed for simplex operation (transmission in one direction at a time) by automatically turning off the sending tone 150 milliseconds after cessation of keying. The sending (transmit) level was set at 0 dBm as measured at the sending end. The receiving level was observed to be around −40 dBm. Keyboard loop current was 20 mA and magnet loop current was 60 mA. The equipment was capable of distortionless signal transmission at rates up to about 270 words per minute (5-level), equivalent to a 200-baud rate. Weitbrecht's half-duplex modem converted five bits of information for letters or figures each typed into two frequencies on the telephone line.

4. Bell Telephone Laboratories demonstrated television by wire in 1927. By 1930, they had an experimental setup of "two-way television" between the AT&T headquarters at 195 Broadway in New York City and the building of the Bell Telephone Laboratories at 463 West Street, two miles away. At Bell Labs, a deaf woman, Evelyn Parry, the national lipreading champion, conversed with her teacher, Marie Pless at the Broadway location. A little later, the deaf dentist Dr. Edwin W. Nies and his wife Maud also conversed over the television-telephone system. Coincidentally, this deaf dentist was later an inspiration to James Marsters; as he approached retirement he offered Marsters his New York City office upon completion of his studies. See "Here at Last! TV-Telephone," 23, for another report on this technology.

5. Robert H. Weitbrecht, demonstration of communication devices at the A. G. Bell Association convention, February 1967.

6. News Release, New York University Office of Information Services, June 30, 1964.

7. Martin L. A. Sternberg, "The Adventures of Martin L. A. Sternberg: Communications Pioneer, Scholar, Actor, and Jaguar Man," *Deaf Life* 8 (February 1996): 26.

8. "American Telephone & Telegraph Board Chairman Shares Interest of the Deaf in Picturephone," *Deaf American* 17 (March 1965): 16.

9. Bell System News Release, October 11, 1965.
10. H. Latham Breunig, "Evolution of the Teleprinter System for Deaf People, 100.
11. TTY call between Robert G. Sanderson and H. Latham Breunig, July 1964.
12. Robert G. Sanderson, "President's Message," *Deaf American*, 18 (October 1965): 30.
13. Robert H. Weitbrecht, "Recent Experiences in the Field of Communication Devices for the Deaf," 3–4.
14. TTY call between Weitbrecht and Marsters, 1965.
15. William Saks, letter to Alexander Graham Bell, March 16, 1922.
16. Alexander Graham Bell, letter to William Saks, March 25, 1922.
17. Andrew Saks, letter to Marsters, February 9, 1965.
18. Andrew Saks, letter to Weitbrecht, February 9, 1965.
19. TTY call between Weitbrecht and Marsters, 1965.

4. The Corporate Windmill

1. Marsters, letter to Saks, February 10, 1965.
2. TTY call between Weitbrecht and Marsters, May 1965.
3. Ibid.
4. W. Schiavoni, letter to Weitbrecht, June 8, 1965.
5. Catherine MacKenzie, *Alexander Graham Bell: The Man Who Conquered Space*, 158.
6. TTY call between Weitbrecht and Merrill Swan, 1965.
7. Weitbrecht, letter to Ray Morrison, August 9, 1965.
8. Saks, letter to Marsters, August 6, 1965.
9. TTY call between Marsters and Saks, August 21, 1965.
10. Ibid.
11. Weitbrecht, letter to Morrison, September 8, 1965.
12. Ibid.
13. Ray E. Morrison, "A New Braille Teleprinter System," *RTTY* 11, no. 12: 2–3.
14. Weitbrecht, letter to Morrison, September 8, 1965.

5. The Frustration Grows

1. Robert G. Sanderson, "President's Message," *Deaf American* 18 (October 1965): 30.
2. Jean Leigh, "Letter to the Editor," *Volta Review* 67 (October 1965): 538.

3. "Helping the Handicapped," *Bell Telephone Magazine* (December 1965): 10.

4. *Communications Act of 1934*, 47 U.S.C.A. 151.

5. "Many Telephone Services Available to the Handicapped," *Bell System News Release* (October 11, 1965): 3.

6. Weitbrecht, letter to H. Latham Breunig, September 28, 1965.

7. J. Kunkler, "V.I.P." *Le Messager* 23 (June–July 1966): 1.

8. Weitbrecht, letter to George W. Fellendorf, September 14, 1966.

9. Marsters, letter to Weitbrecht and Saks, November, 1966.

10. Fellendorf, letter to Marsters, December 7, 1966.

11. TTY call between Weitbrecht and Marsters, November 1966.

12. Marsters, letter to Ralph Larrson, November 8, 1966.

13. Eugene W. Petersen, "Joseph Wiedenmayer: Deaf Diplomat," *Deaf American* 19 (October 1965): 14.

14. Paul L. Taylor, letter to Weitbrecht, November 21, 1966.

15. Weitbrecht, letter to Taylor, November 28, 1966.

16. Ibid.

17. Robert Morrison, letter to Weitbrecht, November 30, 1966.

18. Robert G. Sanderson, "Home Office Notes," *Deaf American* 19 (April 1967): 3.

19. TTY call between Weitbrecht and Marsters, November 1966.

20. Robert H. Weitbrecht, "Teletypewriter Communication over Telephone Lines" (December 12, 1966): 3. Unpublished.

21. TTY call between Weitbrecht and Marsters, December 25, 1966.

22. Ibid.

23. Robert H. Weitbrecht, "Telecommunications in the World of the Deaf" (January 12, 1967): 2. Unpublished.

24. Don G. Pettingill, letter to Marsters, November 28, 1967.

25. Marsters, letter to Pettingill, December 5, 1967.

26. Weitbrecht, letter to Marsters, March 28, 1968.

6. Teletypewriters for the Deaf, Incorporated

1. H. Latham Breunig, letter to Paul L. Taylor, October 1, 1968.

2. Paul L. Taylor, "Establishment of a Large Telephone Communications Network Among the Deaf in St. Louis, Missouri," *Deaf American* 21 (March 1969): 4.

3. Weitbrecht, letter to Ray Smessaert, October 26, 1968.

4. Smessaert, letter to Weitbrecht, November 1, 1968.

5. Weitbrecht, letter to Smessaert, November 7, 1968.

6. Weitbrecht, letter to Marsters, November 27, 1968.

7. Andrea J. Saks, "Phonetype Equals Freedom," *Hearing* 29 (1974): 299.

8. Weitbrecht, letter to Breunig, November 15, 1968.

9. *Dee Cee Eyes* (November 10, 1968).

10. Weitbrecht, letter to Breunig, November 15, 1968.

11. Robert H. Weitbrecht, "Some Noteworthy Events in My Life" (January 14, 1976). Essay.

12. Weitbrecht, letter to Marsters and Saks, March 8, 1969.

13. Weitbrecht, letter to Marsters and Saks, November 18, 1970.

7. Change Agents

1. Weitbrecht, letter to Marsters and Saks, July 5, 1969.

2. Joseph S. Slotnick, "The Telephone-Teleprinter System: Its History, Development and Implications" (paper presented at the Leadership Training Program, Northridge, Calif., February 6, 1969), 4.

3. Weitbrecht, letter to Marsters and Saks, March 8, 1969.

4. Weitbrecht, letter to John Lomax, April 25, 1969.

5. Lomax, letter to Weitbrecht, May 2, 1969.

6. Marsters, letter to Weitbrecht, March 10, 1969.

7. Weitbrecht, letter to Marsters, May 6, 1969.

8. Weitbrecht, letter to Marsters and Saks, May 18, 1969.

9. Weitbrecht, letter to Masters and Saks, May 31, 1969.

10. Weitbrecht, letter to Marsters and Saks, June 4, 1969.

11. Marsters, letter to Weitbrecht, June 18, 1969.

12. Weitbrecht, letter to Marsters, June 17, 1969.

13. Weitbrecht, letter to MITE Corporation, June 20, 1969.

14. TTY call between Lee Brody and H. Latham Breunig, June 24, 1969.

15. Brody, letter to Ray R. Smessaert, June 20, 1969.

16. Weitbrecht, letter to Marsters, July 24, 1969.

17. Weitbrecht, letter to Marsters and Saks, November 17, 1969.

18. Weitbrecht, letter to Marsters and Saks, November 10, 1969.

19. Weitbrecht, letter to Marsters and Saks, November 7, 1969.

20. Leo M. Jacobs, "Of TTYs," *California News* (March 1970): 7.

21. Alfred Sonnenstrahl, e-mail communication to author, August 22, 1997.

22. TTY call between Ed Carney and Weitbrecht, August 27, 1970.

23. St. Louis Telephone-Teletype Communicators' Meeting Minutes, January 26, 1969.

24. TTY users tested the machines by typing "The quick brown fox jumped over the lazy sleeping dog," which contains all the letters of the alphabet. Another sentence used by RTTY amateurs was "Pack my box with five dozen liquor jugs."

25. It is noteworthy that deaf people foresaw the impact of TTY technology on English language development. English is a second language for many deaf people, and they have far fewer primary and incidental opportunities than hearing people to practice English. The result is that there is sometimes a lag in English language development in deaf persons compared to hearing persons of the same age.

26. Weitbrecht, letter to Marsters and Saks, January 5, 1970.

27. AT&T, "Chart of Telephone Equipment for the Totally Deaf," May 7, 1970.

28. "Special Phone Can Be Used by the Deaf and Blind," *IEEE Spectrum* 6 (September 1969): 7.

29. Weitbrecht, letter to Fellendorf, August 5, 1970.

30. Clifton R. Williamson, letter to Breunig, August 14, 1970.

31. Carl Argila, "Telephone Aids for the Deaf—Nationwide Teletypewriter Network for the Deaf," File 38794-43 (May 4, 1970): 1–2.

32. Ibid., 3, 5, 6.

8. The Modem War

1. Reed C. Lawlor, letter to Weitbrecht, August 31, 1970.

2. Lawlor, letter to Weitbrecht, September 28, 1970.

3. TTY call between Marsters and Brody, September 13, 1970.

4. Weitbrecht, letter to Breunig, October 10, 1970.

5. TTY call between Weitbrecht and Marsters, December 6, 1970.

6. Brody, telegram to Weitbrecht, February 28, 1971.

7. Saks, letter to Marsters, July 8, 1969.

8. *GA-SK Newsletter* 2 (February 1971): 1.

9. *First Annual Report of the New England Communications Service for the Deaf, Inc.* (December 1970).

10. Marsters, letter to Ronald S. Callvert, April 20, 1971.

11. Marsters, letter to Callvert, November 18, 1971.

12. All of the papers at this conference were published in the "Green Book," entitled *Proceedings of the First National Conference of Agents of Teletypewriters for the Deaf, Inc., Gallaudet College, Washington, D.C., November 13–14, 1971* (Indianapolis: Teletypewriters for the Deaf, Inc., 1971).

13. J. Thomas Rule, "How the Development of a Modem for the Deaf Has Affected the Course of Deaf People's Lives," in *Proceedings of the First*

National Conference of Agents of Teletypewriters for the Deaf, Inc., Gallaudet College, Washington, D.C., November 13–14, 1971 (Indianapolis: Teletypewriters for the Deaf, Inc., 1971), 71.

14. See "Now Radio Reception for the Deaf," *Deaf American* 23 (July–August 1971): 13.

15. Lawlor, letter to Weitbrecht, September 24, 1971.

16. Weitbrecht, letter to Marsters, February 8, 1972.

17. *GA-SK Newsletter* 3 (July 1972): 1.

18. Jane Miller, letter to Weitbrecht, February 5, 1972.

19. Weitbrecht, letter to Marsters and Saks, August 11, 1972.

20. John R. Pierce, letter to Marsters, September 1, 1972.

21. "Snakes in the Garden of Eden," *GA-SK Newsletter* 3 (November 1972): 3.

22. TDI memo to all authorized agents, March 27, 1973. In many ways, the story of the TTY was like that of Louis Braille's code for blind people—an adventurous innovation that addressed a problem that could be solved but that was subject to competition. In both cases, an assistive device technology was developed by one or more persons with a disability, introducing a major breakthrough in breaking down a communication barrier. The "War of the Modems" was also similar to the "War of the Dots" waged for eighty years in the United States and Britain over the embossed codes for reading and writing by blind people. Indeed, Helen Keller, frustrated with having to use four different embossed codes to read available literature, pleaded for the adoption of one system, which finally occurred in 1932.

23. *GA-SK Newsletter* 3 (July 1972): 4.

24. *Out of the Clatterbox* 1 (November 1969): 6 (Newsletter of the Deaf Telecommunicators of Greater Washington).

9. Foreign Affairs

1. Richard A. Murphy, letter to Marsters, December 21, 1971.

2. Marsters, letter to Murphy, January 3, 1972.

3. Murphy, letter to Marsters, February 7, 1972.

4. TTY call between Saks and Breunig, April 18, 1972.

5. Andrea J. Saks, "Phonetype Equals Freedom," *Hearing* 29 (1974): 300.

6. Philip Timms, letter to A. J. Saks, December 14, 1972.

7. A. J. Saks, letter to Marsters, May 2, 1973.

8. A. J. Saks, letter to British friend C. Mardell, June 11, 1973.

9. A. J. Saks, letter to C. Mardell.

10. Weitbrecht, letter to Marsters, December 19, 1972.

11. Weitbrecht, letter to Marsters, December 18, 1972

12. Marsters, letter to Weitbrecht, December 19, 1972.
13. Weitbrecht, letter to Marsters, December 14, 1972.

10. Revolutions

1. Andrea J. Saks, letter to Ronald V. Dellums, July 8, 1974.
2. A. J. Saks, letter to Andrew Kenyon, January 3, 1974.
3. A. J. Saks, letter to A. C. Roseman, October 25, 1973.
4. Marsters, letter to Monsanto Company, December 5, 1972.
5. Marsters, letter to Joseph E. Wiedenmayer, March 15, 1973.
6. The 1973 Act had a profound effect on the development of new educational and employment opportunities for deaf people. Section 501 prohibited discrimination on the "basis of handicap" in the employment of federal employees. Section 503 required affirmative action plans to employ people with disabilities in any business receiving funds in excess of $2,500 through a federal government contract. For programs and activities receiving federal funding, Section 504 prohibited discrimination against qualified people with disabilities on the "basis of handicap."

The Office of Deafness and Communicative Disorders in the Department of Health, Education and Welfare supported a study to assist vocational rehabilitation personnel in selecting and providing special telecommunications devices for deaf clients. Subsequent amendments to the legislation required the government to implement procedures to ensure that all people, including those who were deaf or blind, could obtain information related to the existence and location of available services. Yet, while the government was expected to provide persons with disabilities with aids, benefits, or services as effective as those provided to others, it had spent $8 billion for its own data processing and telecommunications, which could not be used by deaf citizens who had helped pay for them with their taxes.

7. Alice Hagemeyer started the TTY/Telephone Reference Service at the District of Columbia Public Library. Federal funds available under the Library Services and Construction Act made it possible to purchase the TTY equipment used for this service. Appointed community librarian for the deaf, Hagemeyer subsequently devoted her life to expanding library services around the country for deaf people through the Friends of Libraries for Deaf Action program.

8. Around the country the Pioneers spent hundreds of hours reconditioning and installing TTYs in schools serving deaf students. Other health-related services were also set up in various cities. In Indianapolis, for example, the Telephone Pioneers installed a TTY at the Mental Health Association so that it could expand its services to mentally ill deaf people. This was the only volunteer state mental health organization in the country with such a pro-

gram. Many personal friendships developed among the Telephone Pioneers and the deaf and hearing people with whom they collaborated. In Ohio, Dick Rosenberger single-handedly gave 1,500 hours of his time for TTY reconditioning. He took a course in sign language so that he could communicate more readily with his new deaf friends, obtained TTYs, and set up training classes. When Rosenberger moved to Cincinnati, word spread, and he was soon offering workshops there.

9. Around this time, some people began using the term "Telecommunications Device for the Deaf" (abbreviated "TDD") in a generic manner to include both the older electromechanical teleprinters and the newer electronic devices. Others held an attachment to the term "TTY" and saw no problem with keeping the name as a general, all-inclusive term. The equipment was also not just "for the deaf," but for anyone who wished to communicate with deaf people by telephone.

10. A century earlier, hearing people had expressed this same preference when the voice telephone first came out. The telegraph had had an apparent advantage because of its ability to maintain a permanent record of conversations.

11. Daniel S. Allan and John T. Crandall, *An Evaluation of Telephone Assistance Devices by Deaf and Other Communicatively Disabled Persons* (Menlo Park, Calif.: Stanford Research Institute, August 30, 1974).

12. Robert H. Weitbrecht, "The Telephone-Teletypewriter Communication System for the Deaf," (April 5, 1974), 5–6. Essay.

13. Weitbrecht, letter to Marsters, March 11, 1974.

14. Weitbrecht, letter to Edward C. Merrill, Jr., April 8, 1974. Marsters received an honorary doctoral degree from Rochester Institute of Technology in 1996.

15. TDI's "Red Book," *Teletypewriters Made Easy!*, was compiled by Paul L. Taylor, Frederick N. Stewart, Robert H. Weitbrecht, Dan Skinner, J. Thomas Rule, Eugene L. McDowell, and Thomas R. Schwarz in May 1974.

16. Breunig, letter to A. J. Saks, July 14, 1974.

17. Weitbrecht, letter to A. J. Saks, June 19, 1974.

18. Leona Snyder, letter to Lee Brody, December 28, 1974.

19. Leslie E. Stayer, letter to Brody, February 11, 1975.

20. President Ronald Reagan, letter to Brody, November 16, 1988.

11. Bridges

1. Transatlantic TTY conversation, May 12, 1975.

2. Jack Ashley, tape transcript following transatlantic TTY call, United States Trade Center, London, May 12, 1975.

3. Florian Caligiuri, "The Telephone and the Deaf," 2.

4. The Greater Los Angles Council on the Deaf letter to Pacific Telephone, June 5, 1975.

5. Larry Stewart, cited by Leslie Kafka Halvorsen, "Telecommunications: A Bicentennial Challenge," *Deaf American* 28 (June 1976): 15.

6. Weitbrecht, letter to George W. Fellendorf, September 24, 1976.

7. TTY call between Weitbrecht and Chertok, October 27, 1976.

8. Dan Sawzak, "Today's Service Provisions," In *A Look at Telephone Services for the Handicapped*. (Pacific Telephone and Telegraph, January 14, 1977), 15.

9. *Washington Post*, April 6, 1977, Metro section.

10. U.S. Department of Health, Education and Welfare Office of Human Development, "Social Concerns," In *State White House Conference on the Handicapped Workbook* (Washington, D.C.: U.S. Department of Health, Education and Welfare, 1977), 98.

11. On December 16, 1977, the Robert H. Weitbrecht Company partnership was dissolved.

12. TTY call between Saks and Arnold Daulton, 1978.

13. "Deaf Students Lobby FCC at HEW Birthday Party," *Deaf American* 31 (September 1978): 8.

14. T. Reynolds (Pacific Telephone and Telegraph), letter to Marsters, July 31, 1978.

15. "Deaf Man Criticizes Office of Civil Rights," *Washington Post*, April 12, 1978.

17. Bob Niss, "Deaf Hit NET Rate Structure," *Portland (Maine) Evening Express* (August 10, 1978).

18. TTY call between Weitbrecht and Saks, October 8, 1978.

12. The Changing of the Guard

1. Weitbrecht, communication to Marsters, May 17, 1979.

2. Weitbrecht, letter to Chertok, January 16, 1980.

3. TTY call between Weitbrecht and Marsters, September 13, 1979.

4. Ibid.

5. TTY call between Marsters and Saks, September 16, 1979.

6. Weitbrecht, letter to Chertok, January 16, 1980.

7. Bill SB 597 was signed September 28, 1979, and the program was phased in over a four-year period. It became fully operational on January 1, 1984, and was supported by a surcharge on telephone bills. Judy Tingley and Bill White were leaders in the grassroots deaf community movement for the telecommunications equipment distribution bill in California.

8. When the Royal National Institute for the Deaf established a relay service, the Telephone Exchange for the Deaf, deaf people had trouble ac-

quiring compatible equipment. The service used a V.21 code that was not compatible with Baudot. The Vistel system promoted by the Royal National Institute for the Deaf suffered from design problems, so several other devices were tried, each having problems as well. Ultratec then introduced to England a version of Minicom compatible with both Baudot and V.21, which was used with Typetalk, the relay service that replaced the Telephone Exchange for the Deaf. However, even though Baudot was often used by deaf people in England for regular calls, Typetalk only permitted the V.21 code to be used for incoming emergency calls, not for regular relay calls. Also, although hearing customers of British Telephone were not required to register for use of the common telephone, deaf people were expected to do so for use of Typetalk.

9. See Stephen Von Tetzchner, *Issues in Telecommunication and Disability*, for a discussion of this problem in England. The problem also meant incompatibility with the United States. By 1995, the variety of incompatible telecommunications codes and devices for deaf people remained disheartening. The United States still used the Baudot and ASCII codes. The United Kingdom had Baudot and CCITT V.21, which is not compatible with ASCII. France had Minitel, also called V.23. Switzerland, Poland, Spain, Malta, Germany, Austria, and Italy used EDT, a form of V.21 that was not compatible with England's. Sweden, Norway, and Finland's V.21 was not compatible with other V.21 devices. Denmark and Holland used a telephone keypad called Dual Tone Multi-Frequency (DTMF), which was not compatible with any other telecommunications devices.

In June 1994, after much work by Richard P. Brandt and others, the International Telecommunication Union-Telecommunications Standardization Section (ITU-T Study Group 14) approved a worldwide standard for telecommunications for deaf people. Work toward the V.18 will allow callers anywhere in the world to communicate by TTY. One goal is to have increased access to text telephony in standard terminals manufactured in the future.

10. Daniel Moskowitz, "Technology Is Opening More Jobs for the Deaf," *Business Week* (May 9, 1983): 134.

11. Robert H. Weitbrecht, "Some Noteworthy Events in my Life," (January 14, 1976). Essay.

12. Weitbrecht, letter to Marsters, March 9, 1979.

13. "Accident Victim Still Alive Thanks to Good Samaritan," *San Mateo Times*, June 1, 1983.

13. Legacy

1. Karen Peltz Strauss and Sheila Conlon Mentkowski, letter to Robert Silverstein (with draft of federal bill), January 5, 1988.

2. Paul L. Taylor, letter to Karen Peltz Strauss and Sheila Conlon Mentkowski, January 12, 1988.

3. Remarks of Senator Tom Harkin (Iowa) at the inauguration of President I. King Jordan, Gallaudet University, October 21, 1988.

4. Paul L. Taylor, "Reagan Signs!" GA-SK Newsletter 19 (Fall 1988): 8.

5. Karen Peltz Strauss and Robert E. Richardson, "Breaking Down the Telephone Barrier—Relay Services on the Line," Temple Law Review 64 (1991): 583–607.

6. Draft of bill sent to Silverstein from Peltz Strauss and Conlon Mentkowski, January 5, 1988.

7. The legislative language in Peltz Strauss's draft formed the basis for rules later adopted by the U.S. Department of Justice, requiring TTYs in all 911 centers nationwide.

8. The Americans with Disabilities Act was similar to Section 504 of the Rehabilitation Act of 1973, but it also applied to private businesses and governments not receiving federal funds. It was landmark legislation. The employment section required reasonable accommodations for people with disabilities, including the provision of interpreters and telecommunications devices. Government agencies were expected to provide access to their services, including a direct line for deaf people to 911 telephone networks. The public accommodations section included restaurants, hotels, museums, and similar places, and required the provision of auxiliary aids and services to ensure access. The ADA also required telephone companies to provide dual-party telephone relay services for local and long-distance calls within three years using standards for relay systems established by the FCC.

9. Under Sonnenstrahl's leadership, TDI also advocated for 911 accessibility and consulted with local and long-distance telephone companies, local and state governments, and the FCC. Sonnenstrahl held the office of TDI executive director until his retirement in 1996.

10. Later legislation, the Telecommunications Act of 1996, required that 100 percent of all new television programming be captioned by 2006.

BIBLIOGRAPHY

"A growing network . . . Teletypewriters for the deaf becoming popular." *Deaf American* 20 (May 1968): 3–4.

"A heritage of assistance." *Ohio Bell* (June 1965): 33–35.

Alden, Robert. "New Device allows Deaf to Use Telephone." *Deaf American* 18 (1965): 29.

Allan, Daniel S., Earl J. Craighill, Shmuel S. Oren, Charles L. Jackson, Susan H. Russell, Harold L. Huntley, and Jane Wilson. *A Nationwide Communication System for the Hearing Impaired: Strategies Toward Commercial Implementation.* Stanford, Calif.: Stanford University, 1981.

Allan, Daniel S., and John T. Crandall. *An Evaluation of Telephone Assistance Devices by Deaf and Other Communicatively Disabled Persons.* Menlo Park, Calif.: Stanford Research Institute, 1974.

U.S. House Subcommittee on Telecommunications and Finance. *Americans with Disabilities: Telecommunications Relay Services; Hearing before the Subcommittee on Telecommunications and Finance.* 101st Cong., September 27, 1989.

Banning, W. P. "Bringing Telephone Services to the Deaf." *Bell Telephone Quarterly* (July 1925): 203–10.

Baquis, David. "TDD Relay Services Across the United States." In *Speech to Text: Today and Tomorrow,* ed. Judith E. Harkins and Barbara M. Virvan, 25–43. Gallaudet Research Institute Monograph Series B, no. 2. Washington, D.C.: Gallaudet Research Institute, 1989.

———. "Text Telephone Terminology." *GA-SK Newsletter* 21 (1990): 24.

Basalla, George. *The Evolution of Technology.* New York: Cambridge University Press, 1995.

Bebout, J. Monique. "The Americans with Disabilities Act: American Dream Achieved for the Hearing Impaired?" *Hearing Journal* 43 (June 1990): 11–19.

Bell, Alexander G. "Early Telephony." *Telephone News* 6 (April 1910): 3–7.

———. "Research in Electric Telephony." *Extracts of Proceedings of the Society of Telegraph Engineers.* (1877): *Special General Meeting held on October 31.*

Bellefleur, Philip A. "Radio/Teletype Communications Systems: An Adjunct to Television Captions for the Deaf." *American Annals of the Deaf* 123 (1978): 663–71.

Best, Harry. 1943. *Deafness and the Deaf in the United States.* New York: Macmillan, 1965.

Bijker, Wiebe E., Thomas P. Hughes, and Trevor Pinch. *The Social Construction of Technological Systems.* Cambridge, Mass.: MIT Press, 1993.

Bolling, George H. *AT&T Aftermath of Antitrust: Preserving Positive Command and Control.* Washington, D.C.: National Defense University, 1983.

Boothroyd, Arthur. "Technology and Deafness." *Volta Review* 77 (1975): 27–34.

Bowe, Frank G. "National Survey on Telephone Services and Products." *American Annals of the Deaf* 136 (1991): 278–83.

Brandt, Richard P. "ITU-T Recommendation V.18: The First Communications Standard for the Deaf." In *Report to the Federal Communications Commission on Deaf Telecommunications in Europe,* ed. Andrea Saks. London, 1995.

Breunig, H. Latham. "Evolution of the Teleprinter System for Deaf People in the United States and the Teletypewriters for the Deaf, Inc. (TDI) Organization." In *VII World Congress of the World Federation of the Deaf: Full Citizenship for All Deaf People,* ed. Florence B. Crammatte and Alan B. Crammatte. Silver Spring, Md.: National Association of the Deaf, 1976.

———. "Bell Revisited: Alexander Graham Bell Memorial Lecture." *Volta Review* 76 (1977): 428–34.

"Bringing Telephone Service to the Deaf." *Bell Telephone Quarterly* (July 1925): 203–10.

"British-American Statesmen Linked by TTY." *Deaf American* 28 (1975): 14.

Brooks, John. *Telephone: The First Hundred Years.* New York: Harper and Row, 1976.

Brown, Janet W., and Martha R. Redden. *A Research Agenda on Science and Technology for the Handicapped.* Washington, D.C.: American Association for the Advancement of Science, 1979.

Brown, Ruth S. "Learning with TDDs." *Perspectives for Teachers of the Hearing Impaired* 3 (November/December 1984): 4–6.

Bruce, Robert V. *Bell: Alexander Graham Bell and the Conquest of Solitude.* Boston: Little, Brown, 1973.

Bubeck, Allan F. "How to Use the Telephone in Emergencies." *Silent Worker* 13 (1961): 23.

Burlingame, Roger. *Out of Silence into Sound: The Life of Alexander Graham Bell.* New York: Macmillan, 1964.

Cagle, Sharon J., and Keith M. Cagle. *GA and SK Etiquette: Guidelines for Telecommunications in the Deaf Community.* Bowling Green, Ohio: Bowling Green Press, 1991.

Caligiuri, Florian A. "The Telephone and the Deaf." *Journal of Rehabilitation of the Deaf* 8 (1974): 1–3.

Cardwell, Donald. *The Norton History of Technology.* New York: W. W. Norton, 1995.

Castle, Diane L. "Telecommunications Technologies Used by Employees Who Are Deaf." *JADARA* 27 (1994): 1–8.

Cerf, Vinton. "The Electronic Mailbox: A New Communication Tool for the Hearing Impaired." *American Annals of the Deaf* 123 (1978): 768–72.

Cherry, Colin. "Invention and 'Revolution.'" In *The Social Impact of the Telephone*, ed. Ithiel De Sola Pool. Cambridge, Mass.: MIT Press, 1977.

Coe, Lewis. *The Telegraph: A History of Morse's Invention and Its Predecessors in the United States.* Jefferson, N.C.: McFarland and Co., 1993.

Cohen, I. Bernard. *Revolution in Science.* Cambridge, Mass.: Harvard University Press, 1985.

Conlon-Mentkowski, Sheila. "Overview of State-Regulated Relay Services." In *Speech to Text: Today and Tomorrow*, ed. Judith E. Harkins and Barbara M. Virvan, 19–23. Gallaudet Research Institute Monograph Series B, no. 2. Washington, D.C.: Gallaudet Research Institute, 1989.

Cooley, Mike. *Architect or Bee? The Human/Technology Relationship.* Boston: South End Press, 1980.

Craighill, Earl J. *Research in Telecommunications for the Deaf.* Menlo Park, Calif.: Stanford Research Institute, 1979.

Crammatte, Alan. B. *Deaf Persons in Professional Employment.* Springfield, Ill.: Charles C. Thomas, 1968.

Crandall, John T. *Summary of a Study of Several Telephone Communication Assistance Devices for Deaf People.* Washington, D.C.: Office of Deafness and Communication Disorders, Rehabilitation Services Administration, Social and Rehabilitation Services, Department of Health, Education and Welfare, August 1974.

"Deaf Students Lobby FCC at HEW Birthday Party." *Deaf American* 31 (1978): 7.

DeLand, Fred. *Dumb No Longer: Romance of the Telephone.* Washington, D.C.: Volta Bureau, 1908.

Derry, Thomas K., and Trevor I. Williams. *A Short History of Technology from the Earliest Times to A.D. 1900.* New York: Dover, 1960.

de Sola Pool, Ithiel, ed. *The Social Impact of the Telephone.* Cambridge, Mass.: MIT Press, 1977.

Dicker, Leo. "Suggested Procedures for Using Telecommunication Services." *Deaf American* 27 (1974): 9–10.

Dilts, Marion M. *The Telephone in a Changing World.* New York: Longmans, Green and Co., 1941.

Drexler, K. Eric, Chris Peterson, and Gayle Pergamit. *Unbounding the Future: The Nanotechnology Revolution.* New York: Quill, 1991.

Driver, Sherry, Terrance Easterwood, and John Schaub. "Development of a High School TTY Program." *Volta Review* 81 (December 1979): 517–20.

DuBow, Sy, Larry Goldberg, Sarah Geer, Elaine Gardner, Andrew Penn, Sheila Conlon, and Marc Charmatz. *Legal Rights of Hearing-Impaired People.* Washington, D.C.: Gallaudet College Press, 1984.

Emerton, R. Greg, Susan Foster, and Hariette Royer. "The Impact of Changing Technology on the Employment of a Group of Older Deaf Workers." *Journal of Rehabilitation of the Deaf* 21 (October 1987): 6–18.

Erber, Norman P. *Telephone Communication and Hearing Impairment.* San Diego: College-Hill Press, 1985.

Faulhaber, Gerald R. *Telecommunications in Turmoil: Technology and Public Policy.* Cambridge, Mass.: Ballinger, 1987.

Fellendorf, George W. "The Birth of the TDD: I was There." *Austin News* (Winter 1990): 17, 20.

Flanagan, James L. "New Approaches in Telephone Use by the Deaf." In *Accent on Unity: Horizons on Deafness. Proceedings of National Forum I,* ed. Harriet G. Kopp. Washington, D.C.: Council of Organizations Serving the Deaf, 1968.

Florman, Samuel C. *Blaming Technology: The Irrational Search for Scapegoats.* New York: St. Martin's Press, 1981.

Gabler, Edwin. *The American Telegrapher: A Social History, 1860–1900.* New Brunswick, N.J.: Rutgers University Press, 1988.

Gallaudet College. *Guide for TTY-Telephone Communication.* Washington, D.C.: Gallaudet College Public Service Programs, 1974.

Gandy, Oscar H., Jr., Yvonne Matthews, Herma Williams, and Gladys Vaughn. "Consumer Education and the TTY: A Progress Report." *Volta Review* 83 (December 1981): 477–79.

Geoffrion, Leo D. "An Analysis of Teletype Conversations." *American Annals of the Deaf* 127 (1982): 747–52.

Glaser, Robert E. "Telephone Communication for the Deaf." *American Annals of the Deaf* 127 (1982): 550–55.

Graham, Bill. *One Thing Led to the Next: The Real History of TTYs.* Evanston, Ill.: Mosquito Publishing, 1988.

Gutshall, Ruth. "Telephone Communication Devices for the Deaf." *Stanford Research Institute Intercom* no. 68 (August 23, 1967).

Heil, Joseph. "Disabled People Can Be an Important Market Segment: A Large-Business Perspective." In *Papers Presented at the Annenberg Washington Program and the Gallaudet Research Institute Joint Forum on Marketplace Problems in Communication Technology for Disabled People,* February 20–21. Washington, D.C.: Annenberg Washington Program in Communication Policy Studies of Northwestern University, 1986.

Hellstrom, Gunnar. "International Standardization of Text Telephony." In *Report to the Federal Communications Commission on Deaf Telecommunications in Europe,* ed. Andrea Saks. London, 1995.

"Helping the Handicapped." *Telephone News* (December 1965): 8–11.

"Here at Last! TV-Telephone." *Silent Worker* 9 (1957): 23.

Holmes, Pamela. "Meet the Challenge: Create a More Accessible Public Telephone Environment." *GA-SK Newsletter* 24 (1993): 22–23.

Holzmann, Gerard J., and Björn Pehrson. *The Early History of Data Networks.* Los Alamitos, Calif.: IEEE Computer Society Press, 1995.

Irgens, Henning F. C. "Telephone Communication Experiments." In *Report of the 42d Meeting of the Convention of American Instructors of the Deaf,* 55–56. CAID, 1965.

Ives, Herbert E. "Two-Way Television." In *Two-Way Television and a Pictorial Account of its Background,* 23–40. New York: Bell Telephone Laboratories, 1930.

Jackson, M. T. "Long Lines: The Increasing Public Use of Long Distance Telephone Facilities Adds to Western Electric Responsibilities." *Western Electric Magazine* (May 1929): 3–6.

Jamison, Steven L., and John T. Crandall. "Telephonic Communication Assistance Devices for Deaf People." *Deafness Annual* 4 (1974): 143–58.

Jensema, Carl J. "Telecommunications for the Deaf: Echoes of the Past—A Glimpse at the Future." *American Annals of the Deaf* 139 (1994): 22–27.

Jesperson, James. "Overview of Telecommunications Devices for Deaf Persons, Exclusive of Television." In *VII World Congress of the World Federation of the Deaf: Full Citizenship for All Deaf People,* ed. Florence B. Crammatte and Alan B. Crammatte. Silver Spring, Md.: National Association of the Deaf, 1976.

Jewett, F. B. "Science and Telephony." *Western Electric Magazine* (March 1929): 3–6.

Johnson, Harold A., and Lyle E. Barton. "TDD Conversations: A Context for Language Sampling and Analysis." *American Annals of the Deaf* 133 (1988): 19–25.

Jones, J. Paul. "Telecommunication and the Deaf." *Deaf American* 39 (1989): 19–20.

Jones, Ray L. *Telephone Communication for the Deaf: Speech Indicators Manual.* Northridge, Calif.: San Fernando Valley State College, 1970.

Jurgen, Ronald K. "Devices for the Disabled: A Production Dilemma." *IEEE Spectrum* 13 (1976): 45–49.

Kingman, W. S. "Independent Telephone Technical Advances." Paper presented to a conference of daily press and security analysts. Washington, D.C., February 1965.

Kraus, Constantine R., and Alfred W. Duerig. *The Rape of Ma Bell: The Criminal Wrecking of the Best Telephone System in the World.* Secaucus, N.J.: Lyle Stuart, 1988.

Lambert, T. "S. Africa Tackling Erratic Phone System." *Los Angeles Times,* March 1975.

Lang, Harry G. *Silence of the Spheres: The Deaf Experience in the History of Science.* Westport, Conn.: Bergin and Garvey, 1994.

Lang, Harry G., and Bonnie Meath-Lang. *Deaf Persons in the Arts and Sciences: A Biographical Dictionary.* Westport, Conn.: Greenwood Press, 1995.

Lunde, Anders S., and Stanley K. Bigman. *Occupational Conditions Among the Deaf: A Report on a National Survey Conducted by Gallaudet College and the National Association of the Deaf.* Washington, D.C.: Gallaudet College, 1959.

MacKenzie, Catherine. *Alexander Graham Bell: The Man Who Conquered Space.* Boston: Houghton Mifflin, 1928.

MacQuivey, Donald R. "Use of Customer-Owned Equipment on Public Telephone Lines by Deaf Persons." Menlo Park, Calif.: Stanford Research Institute, 1967.

Mann, Willis J. " 'Hotline for the Deaf' Established in D.C. Area." *Deaf American* 24 (November 1971): 13–14.

Marsters, James C. "Radio-Teletypewriters for the Deaf—Present and Future." In *Proceedings of the First National Conference of Agents of Teletypewriters for the Deaf, Inc., Gallaudet College, Washington, D.C., November 13–14, 1971.* Indianapolis: Teletypewriters for the Deaf, 1971.

Marzano, Rudy. "But What Have You Done for Us Lately?" *Bell Telephone Magazine* 59 (1980): 35–41.

Meadow, Lloyd. "Implications of Technological Changes for Employment of Deaf Persons." In *New Vistas for Competitive Employment of Deaf Persons,* ed. William N. Craig and James L. Collins, 63–73. Readings in Deafness Monograph No. 2. Knoxville: Journal of Rehabilitation of the Deaf, 1970.

Moore, Omar K. "A Warm Medium of Communication." *Bell Telephone Magazine* (1972): 30–32.

———. "The Inclusion of the Deaf within Communications Networks." *American Annals of the Deaf* 119 (1974): 597–601.

Office of Human Development. *Social Concerns. State White House Conference Workbook.* Washington, D.C.: U.S. Department of Health, Education and Welfare, 1977.

Office of Technology Assessment. *Technology and Handicapped People.* Washington, D.C.: U.S. Government Printing Office, 1982.

Olesen, K. G. "Survey of Text Telephones and Relay Services in Europe: Final Report." In *Information Technologies and Sciences: Proceedings of COST 219 Conference.* Luxembourg: Office for Official Publications of the European Communities, 1992.

Pacey, Arnold. *The Culture of Technology.* Cambridge, Mass.: MIT Press, 1994.

Pacific Telephone and Telegraph. "A Look at Telephone Services for the Handicapped: First report." January 1977.

Pflaum, Michael E. "The California Connection: Interfacing a Telecommunication Device for the Deaf (TDD) and an Apple Computer." *American Annals of the Deaf* 127 (1982): 573–84.

Pound, Arthur. *The Telephone Idea: Fifty Years After.* New York: Greenberg, 1926.

Romney, Frederic C. "Deaf Students Use the Telephone for the First Time." *Volta Review* 77 (1975): 125–28.

Rule, J. Thomas, Jr. "How the Development of a Modem for the Deaf Has Affected the Course of the Deaf People's Lives." In *Proceedings of the First National Conference of Agents of Teletypewriters for the Deaf, Inc., Gallaudet College, Washington, D.C., November 13–14, 1971.* Indianapolis: Teletypewriters for the Deaf, 1971.

Rule, J. Thomas, Jr., and Robert C. Sampson. "First Regional TTY Workshop and Conference." *Deaf American* 22 (1970): 19–20.

Saks, Andrea J. "Phonetype Equals Freedom." *Hearing* 29 (1974): 299–302.

———. "Comments on the Regulation of British Telecom's Prices." Report presented to the Office of Telecommunications, London, March 1992. Photocopy.

———. "Comment to the Office of Secretary." In *Federal Communications Commission: IAD Document No. 93–02, Assignment of N–1–1 Codes to Facilitate Access to Telecommunications Relay Services.* Washington, D.C.: FCC, 1993.

———. "World TTY Compatibility at Hand." *Silent News* (June 1994): 32.

———. "Report to the Federal Communications Commission of the United

States of America on Deaf Telecommunications in Europe." 1995. Photo-copy.

Saks, Andrew. "Commercial Telephone Answering Service and Problems in Serving Deaf Subscribers." In *Proceedings of the First Conference of Agents of Teletypewriters for the Deaf, Inc., Gallaudet College, Washington, D.C., November 13–14, 1971.* Indianapolis: Teletypewriters for the Deaf, 1971.

Sampson, Robert C. "An Hour at 393 Seventh Avenue." *Deaf American* 21 (1969): 11.

Scadden, Lawrence. "Communication Technology for Disabled People: Problems in the Marketplace." In *Papers Presented at the Annenberg Washington Program and the Gallaudet Research Institute Forum on Marketplace Problems in Communications Technology for Disabled People.* Washington, D.C.: Annenberg Washington Program in Communication Policy Studies of Northwestern University, 1986.

Schultz, Stephen. "The History and Development at the California Relay Service." *Volta Review* 92 (1990): 93–97.

Scribner, C. E. "Log-Fire Reminiscences of a Pioneer. Bell's Invention of the Telephone and Its Introduction to the Public." *Western Electric Magazine* (March 1926): 3–6.

Shiers, George, ed. *The Electric Telegraph: An Historical Anthology.* New York: Arno Press, 1977.

Slotnick, Joseph S. "The Telephone-Teleprinter System: Its History, Development and Implications." Paper presented at the Leadership Training Program in the Area of the Deaf, Northridge, Calif., San Fernando Valley State College, February 1969.

Smith, Gale M. "Teletype Service for the Totally Deaf." *Volta Review* 65 (1963): 579–83.

———. "What Hath God Wrought." *Volta Review* 67 (1965): 505–7.

Smith, Merritt Roe, and Leo Marx, eds. *Does Technology Drive History? The Dilemma of Technological Determinism.* Cambridge, Mass.: MIT Press, 1994.

Stanford Research Institute. "Summary of Planning Meeting Concerned with Communications with the Deaf." Menlo Park, Calif.: Stanford Research Institute, January 1967.

Stern, Virginia W., and Martha R. Redden. "Selected Telecommunications Devices for Hearing-Impaired Persons. Background Paper #2." In *Technology and Handicapped People.* Washington, D.C.: Office of Technology Assessment, December 1982.

Stevenson, Leslie, and Henry Byerly. *The Many Faces of Science: An Introduction to Scientists, Values, and Society.* San Francisco: Westview Press, 1982.

Stoffels, Robert. "Telecommunications Devices for the Deaf." *Telephone Engineer and Management* (October 1980): 69–73.

Stone, Alan. *Wrong Number: The Breakup of AT&T.* New York: Basic Books, 1989.

Stovall, Margaret. "Teletype Breakthrough for Deaf." *Pasadena (California) Star News* November 1964, 3.

Strassler, Barry. "Telecom and You." *Deaf American* 32 (1979): 29.

———. "The ASCII-Baudot Dilemma. Telecom and You." *Silent News* (March 1980): 18.

Strauss, Karen Peltz. "Telecommunications Issues for Disabled Persons: The Role of Federal and State Regulation." In *Papers Presented at the Annenberg Washington Program and the Gallaudet Research Institute Forum on Marketplace Problems in Communications Technology for Disabled People.* Washington, D.C.: Annenberg Washington Program in Communication Policy Studies of Northwestern University, 1986.

———. "Implementing the Telecommunications Provisions." In *The Americans with Disabilities Act: From Policy to Practice,* ed. Jane West, 238–67. New York: Milbank Memorial Fund, 1991.

———. "Universal Service: What Is It and How Can I Get It?" *NAD Broadcaster,* June 9, 1996.

"Talk, Hear, See on This Phone: Two-Way Television Is Demonstrated in Laboratory as an Engineering Stunt." *Popular Science Monthly* 22 (July 1930): 123.

Tanenbaum, Sandra J. *Engineering Disability.* Philadelphia: Temple University Press, 1986.

Taylor, Paul L. "Establishment of a Large Telephone Communications Network Among the Deaf in St. Louis, Missouri." 1969. Photocopy.

———. "Establishment of a Large Telephone Communications Network Among the Deaf in St. Louis, Missouri." *Deaf American* 21 (1969): 3–5.

———. "Weather Report Service for the Deaf Successfully Implemented." *Deaf American* 21 (June 1969): 29.

———. "Telephone Relay Service: Rationale and Overview." In *Speech to Text: Today and Tomorrow,* ed. Judith E. Harkins and Barbara M. Virvan, 11–18. Gallaudet Research Institute Monograph Series B, no. 2. Washington, D.C.: Gallaudet Research Institute, 1989.

Telecommunications Access With and Within the Federal Government: A Consideration of Issues and Applications for Telecommunication Devices for Deaf Persons (TDDs). Final Report. Washington, D.C.: U.S. Architectural and Transportation Barriers Compliance Board, August 3, 1984.

Telecommunications for the Deaf, Inc. "Telecommunication Devices and Their Role in Deaf Society." *Deaf American* 31 (May 1979): 22.

"Telephone Services Available to the Handicapped." *Deaf American* 18 (1965): 8, 10.

"Teletypewriters for the Deaf Becoming Popular." *Deaf American* 20 (1968): 9.

Thornton, Patricia. "Communications Technology—Empowerment or Disempowerment?" *Disability, Handicap and Society* 8 (1993): 339–49.

"TTY Call Connects Italy and the United States." *Deaf American* 28 (May 1976): 7, 9.

Tucker, Durward J. *RTTY From A to Z.* Port Washington, N.Y.: Cowan, 1970.

U.S. Congress. "Background Paper #2: Selected Telecommunications Devices for Hearing Impaired Persons (OTA-BP-H-16)." Washington, D.C.: Office of Technology Assessment, 1970.

von Tetzchner, Stephen, ed. *Issues in Telecommunication and Disability.* Luxembourg: Office for Official Publications of the European Communities, 1991.

Ward, Paul R. "Telephone Communication for Deaf People: Hearing Research Project." University of Exeter, Institute of Biometry and Community Medicine, February 1974.

Wasserman, Neil H. *From Invention to Innovation: Long-Distance Telephone Transmission at the Turn of the Century.* Baltimore, Md.: Johns Hopkins University Press, 1985.

Webster, Andrew. *Science, Technology, and Society: New Directions.* New Brunswick, N.J.: Rutgers University Press, 1991.

Whitney, L. Holland. "Services for Special Needs." *Bell Telephone Magazine* (Summer 1965): 25–32.

Wiedenmayer, Joseph. "The Deaf, the Blind and the Telephone." *Deaf American* 24 (1972): 9.

Wyman, Raymond, and Todd Eachus. *A Field Test of Electronic Telecommunication Terminals for the Deaf.* Amherst, Mass.: University of Massachusetts, 1974.

Papers by Robert H. Weitbrecht

"Current Integrator for Astronomical Photoelectric Photometry." *Review of Scientific Instruments* 28, no. 11 (November 1957): 883–88.

"Recent Experiences in the Field of Communication Devices for the Deaf." Report to the Alexander Graham Bell Association for the Deaf, February 18, 1965.

"Teletypewriting Over Telephone Lines." Menlo Park, Calif.: Robert H. Weitbrecht Co., November 12, 1965.

"Instructions for the Use of Teleprinter Equipment Over the Telephone

Line." *Technical Bulletin* no. 1 (April 1966): Menlo Park, Calif.: Robert H. Weitbrecht Co.

"Carrier-Current Signaling System (Sound-Activated Transmitter Unit)." Menlo Park, Calif.: Applied Communications Corp., October 7, 1966.

"Telephone-Teletype Terminal Unit (Acoustic Coupler)." Menlo Park, Calif.: Robert H. Weitbrecht Co., October 7, 1966.

"Teletypewriter Communication over Telephone Lines." Menlo Park, Calif.: Robert H. Weitbrecht Co., December 12, 1966.

"Telecommunications in the World of the Deaf." Paper presented at a conference at Stanford Research Institute, Menlo Park, Calif., January 12, 1967.

"Teletypewriter Machine Specifications for a Teletype System for the Deaf over the Regular Telephone System." Menlo Park, Calif.: Robert H. Weitbrecht Co., January 16, 1967.

"Physical Description of Terminal Unit: Phone-TTY System." *Technical Bulletin* no. 2 (April 1967). Menlo Park, Calif.: Robert H. Weitbrecht Co.

"Teleprinter Machine Connections for Use with Phone-TTY System." *Technical Bulletin* no. 3 (May 1967). Menlo Park, Calif.: Robert H. Weitbrecht Co.

"Testing a Teletypewriter Machine." *Technical Bulletin* no. 4 (July 1967). Menlo Park, Calif.: Robert H. Weitbrecht Co.

"Telephone Handset Check and Adjustment." *Technical Bulletin* no. 5 (November 1967). Menlo Park, Calif.: Robert H. Weitbrecht Co.

"Adjustments in the Range Finder-Selector Magnet Area." *Technical Bulletin* no. 6 (November 1967). Menlo Park, Calif.: Robert H. Weitbrecht Co.

"Ordering Telephone Service for Phonetype Operations." *Technical Bulletin* no. 7 (March 1968). Menlo Park, Calif.: Applied Communications Corp.

"Teletype Model 15 Machines [Including Model 19]. *Technical Bulletin* no. 8 (August 1968). Menlo Park, Calif.: Applied Communications Corp.

"Telephone-ring Signaling Unit Model SU-1." Menlo Park, Calif.: Applied Communications Corp., January 1969.

"Automatic Electric Co. Telephones." Menlo Park, Calif.: Applied Communications Corp., September 9, 1969.

"Message Control Unit (MCU) for ACU-Phonetype System." *Technical Bulletin* (January 6, 1970). Menlo Park, Calif.: Applied Communications Corp.

"Message Control Unit. Technical Memo (revised from September 3, 1970)." Menlo Park, Calif.: Applied Communications Corp., January 1971.

"Technical Specifications; Phonetype: A Telephone-Teletypewriter System for the Deaf." Menlo Park, Calif.: Applied Communications Corp., February 1971.

"A Telephone-Teletypewriter System." *Proceedings of the First Conference of Agents of Teletypewriters for the Deaf, Inc., Gallaudet College, Washington, D.C., November 13–14.* Indianapolis: Teletypewriters for the Deaf, 1971.

"All-Acoustic Coupling Cradle Box (for Phonetype I to Work with General Telephone Co. Units)." Menlo Park, Calif.: Applied Communications Corp., August 10, 1971.

"Two Kinds of Teleprinters." Menlo Park, Calif.: Applied Communications Corp., September 9, 1971.

"Comparison—PHONETYPE System with Other Data Systems." Menlo Park, Calif.: Applied Communications Corp., September 10, 1971.

"Studies by Robert H. Weitbrecht." Menlo Park, Calif.: Applied Communications Corp., November 8, 1971.

"Operating Instructions: Phonetype IV (and III) Terminal Unit." Menlo Park, Calif.: Applied Communications Corp., February 23, 1972.

"Compatibility Is a Two-Way Relation." Menlo Park, Calif.: Applied Communications Corp., June 23, 1972.

"My Days as an Astronomy Student at the University of California, Berkeley, Calif., 1940–1942." April 22, 1973. Essay.

"L-Type Earpiece Element in New Bell System 500-Series Telephone Sets." Menlo Park, Calif.: Applied Communications Corp., April 30, 1973.

"Phonetype 8 Terminal Units for England." Menlo Park, Calif.: Applied Communications Corp., June 23, 1973.

"The Telephone-Teletypewriter Communication System for the Deaf." Menlo Park, Calif.: Applied Communications Corp., April 5, 1974.

"Model 15 Keyboards—Two Kinds: 7.00 Unit and 7.42 Unit. Menlo Park, Calif.: Applied Communications Corp., 1974.

"Summary in Regard to Model 15 TTY Keyboards." Menlo Park, Calif.: Applied Communications Corp., 1974.

"Some Noteworthy Events in My Life." January 14, 1976. Essay.

"It All Came From Mount Lassen, or How Did the Deaf People Get Their Telephone System?" February 1976. Essay.

"Radio Broadcasting of "Phonetype" Signals." Menlo Park, Calif.: Applied Communications Corp., March 15, 1976.

"Teletypewriter Range Finder Considerations." Menlo Park, Calif.: Applied Communications Corp., n.d.

INDEX

Page numbers in italics indicate photographs. All subentries for *Saks* refer to Andrew Saks, unless otherwise noted.